SCIENTIFIC IMPLICATIONS:
Fantasy, Fancy, and Reality

I0475324

Copyright 2010 Austin P. Torney

austintorn@aol.com

Sometimes the mind so much wants to do it function
To know all, that it speculates its way to 'truth',
Not realizing that its mere pronouncement
Just floats in the thin air as an unsupported belief.

WORDS VERSUS ILLUSTRATION

The writer's pen stood forth, being first,
Instructing the artist's stylus
To illustrate the words of the epic,
Noting that a picture was worth a thousand words.

"Perhaps we don't even need the words",
Retorted the artist's stylus,
"As I am worth so many".

"Well," replied the writer's pen,
"It's true that many people now refuse
To read books without lots of pictures in them."

"How sad, for I guess some words
Are needed to round out the tale."

"True, for the two sides of the brain
Can then combine in unity."

"Or I could draw the pictures first
And then you could write the words."

"It could be like that sometimes, I suppose."

"OK, shake; it's a deal either way,
For we need each other."

TO THE END(S) OF THE UNIVERSE

I took a road trip
Through the universe recently,
Smoking some pot
And playing the radio loud.

Holy-moly, there's nothing holy out there.
In fact, it's a very uncongenial place for life.
I'd much rather be in Australia

96% of it was useless dark energy and dark matter;
The rest was mostly rocks gases and dust;
Dangerous radiation zapped all over the place;
And it was fricken freezing!

Oh, what I would have given to be in Canada.

Whatever designed the universe
Certainly didn't have life in mind;
It even took evolution billions of years
To fine-tune us to the earth.

Then we nearly got wiped out
By huge disasters right and left,
Even once shrinking back down
To a population of around 2000.

I saw the graveyards of the stars
And some stellar nurseries, too;
All kinds of energy swirled about—
When it wasn't exploding and wreaking havoc.

I stopped to eat at the restaurant
At the end of the universe,
On a moon,
But, it had no atmosphere,
Plus, all the food had been microwaved,
By the CMBR.

What a wasteland of a wilderness of wilds
Of a whole bunch of crap
That nearly went on forever in every direction;
This was as much of a place
Unsuited for life that there ever could be.

I'm back, thank my lucky stars,
Noting that, 14 billion years
After the initial chaos, here we are,
Having beaten the odds.

Well, someone had to!
We won the universal lottery jackpot.

Oh cripes, here comes a humongous asteroid!
Darn, all that luck for nothing.
Double '00' has come up.

It was only a matter of time.

NO TIME MACHINES

Now, tongue in cheek,
I'll tell why there are no time machines.

It isn't that no one ever came back
From the future to see us,
Although that is still a good reason
For no time machines being possible.

Nor is it that there can't be
A future going on somewhere ahead of time,
As that's a fine one, too.

It is that women prevented time machines
From being invented;
For, every time a man said,
"Honey, I'm going out to the garage
To work on my time machine"
The woman in his life would reply
"That's impossible, dear.
Stop wasting your time;
There is housework to be done
And grass to be cut."

The man would still sneak out
To try to work on his time machine,
But the woman would find him
And, once again, say something like,
"That's impossible, you nut head.
Get in here and do something useful!"

And that's why there are no time machines!

HENRY CAVENDISH

He was a century ahead of his time,
On electrical conductivity, for example,
But kept everything so secret
That much of it didn't come out
Until the century had passed.

Without telling any one,
Cavendish discovered or anticipated
The law of conservation of energy,
Ohm's law, Dalton's Law of Partial Pressures,
Ricther's Law of Reciprocal Proportions,
Charles's Law of Gases,
And the principles of electrical conductivity.

And that's just some of it;
There was also tidal friction,
Atmospheric cooling, freezing mixtures,
Heterogeneous equilibria,
Clues to the noble gases,
And of course the gravitational constant,
And the weight of the Earth—
From a nautilus looking machine of weight,
Counterweights, pendulums,
Shafts, and torsion wires.

He was so shy that it was known
That on no account was he
To be approached or even looked at.

Those who sought his views
Were advised to wander into his vicinity,
As if by accident,
And, to "talk as it were into vacancy".

If their remarks were not worthy,
An actual vacancy would quickly appear.

WINTER SNOW

I hope the snow keeps up,
Though not to bury us all around,
But only so that it doesn't come down.

AN IDEA WITH NO HAIR

I read something,
But I lost the magazine;
So I will try to reconstruct it...

We were once very furry,
But now have very little hair
On our bodies.

Most animals still have
A lot of hair or fur,
But for the larger beasts
That ever need to cool
A large mass.

It's warmer now,
So even the naked elephant
Is no longer a Wooly Mammoth.

When eats and treats were abundant,
Our furry ancestors could just
Laze around, not exerting themselves;

But, when forest shrank,
'We' had to walk/run
Over long distances
To secure our prey and food,
And so some thick clogged hair
All over would not do so well for us.

Hair was thus selected out,
In conjunction with
The arrival of sweat glands
That could dump
12 liters a day.

We were pink underneath our fur,
This turning to black in the hot regions.

Hair on the head still works
To ward off the direct sun.

Via our new epidermal sweat glands,
We could increase
Our body and brain size
And exert ourselves more.

TOP SECRET COSMOLOGICAL CONJECTURE

1.
Energy conservation is absolute in each O region.
Energy conservation globally is undefined
Since there is no time translation globally.

2.
Inflation is the ultimate free lunch (almost)
Which drives each O region to a flat metric
Which locks each O region
Within the Bekenstein bound
And creates a zero energy Universe.
(Almost)

3.
The process of inflation required
A given amount of seed energy to it.
Once inflation is launched
The work function of expansion
Is used to create all the mass energy in the Universe
Without violating the conservation of energy.

4.
The small seed energy to launch inflation
Comes from a Quantum Fluctuation
Which creates both
A negative energy time reversed inflating Universe
And a positive energy time forward inflating Universe,
The sign of energy and arrow of time being always forward
And the energy always positive
From the perspective of any observer.

This is a process analogous to particle pair creation
And conforms to numerous cosmological models:
Sakharov multisheet model,
The no boundary model,
The Venzianio and Gasperini
String cosmological model etc.

5.
The end state of the Universe
Will be an empty De Sitter State
With a constant vacuum energy density.
This De Sitter space will be the birthing ground
For baby BIVERSES as described by a model
Like the Carroll Chen proposal.

Please keep this under your hat—in your mind!

In the same way as particles
Emerging from the vacuum
Are produced in pairs and characterized
By opposite physical properties
(i.e., opposite charges,
Momentum, angular momentum, etc.)
To avoid violations of conservation laws,
The Universes are created
Are also produced in pairs,
And are characterized
By opposite kinematic properties.

One of the Universes expands
While the other one shrinks.
However, the shrinking Universe
Behaves as if it was
Traveling backwards in time
With respect to the coordinates
Playing the role of time in mini superspace.

IN MODERATION

Graybeard is lively and fun,
Being loved by everyone;
He points me to readings wide
That on the internet reside.

He, too, is a brilliant one...

But I say too much here with this poetry,
For he's well out there for all to see.

Now that he is both a moderator
And a poster, like a schizorater,
When we ask "How are you?",
He says "OK"... "and so am I, as well."

FINITE EXISTENCE

To infinity existence cannot be,
As all here is not of great density,
But of a finite amount
From which is formed our account.

THE PRESENCE

Some have such an experience of a presence
That they then leap to call the experience
God, the cosmos, or the All.

The experience is indeed felt,
But is something else entirely.

It is the left brain,
Wherefrom the sense of self comes,
Sensing the unspeaking right brain
As a presence.

To call it God/All is of a cerebral mistake,
Not to mention it being a giant leap
Of unfounded opinion rather than fact.

You can even feel it a bit now,
Kind of, however slight,
Such as when having dinner,
Driving, or doing some task.

It then feels like your left brain
Is full, of course,
And that the right is empty,
The left side even feeling warmer;
But, pause from the task for a second
And holistically observe something as a whole,
Such as a furniture cabinet or a car,
Not getting into the details.

Now the right brain is the half
That feels warm and full—
And maybe even a little spacey.

That's all it is, folks.

In some it can get very extreme,
Medically bothersome even,
Like having two selves
Or positing an intruding self
That is not one's own.

Patients whose hemispheres' connection
Is severed to relieve epilepsy may even develop
Two separate consciousnesses,
Often at odds with each other,
Such as getting choked with one hand,
The other hand trying to take it away.

Another, differently sourced feeling,
Of being One with the Cosmos
Comes during meditation,
But is only of the calming
Of the brain neurons
In the parietal lobe,
Those that maintain
The identification of the self
And also of where the body ends
And the rest of the universe begins.

That's all it is.

This can also happen
During praying or chanting;
It's been measured in monks.

Who then, upon recalling,
Or during a state of meditation,
Or just by living,
Can use an experience
Of doubtful analysis
By the mere state of being
To say anything further
About the true nature,
Why, source, how, and wherefore
Of the conscious awareness
Of felt sensation—
Without consideration of the
Electrochemical states beneath?

There are those who feel
They have to say,
Based on introspection alone,
And those who do not,
For science also informs them
Of the states beneath.

OUT THERE?

The mind-brain projects the scene
To appear out whence its E/M waves came,
Utilizing the clues of photons' angles,
Source of direction, intensity, and distance, etc.;
But, what you really actually see
Is only the inside of your head.

THE STAGE IS THE REAL PLACE
OF THE REAL EVENTS OF THE WORLD

One's consciousness witnesses and views it.

It is the audience:
It can take a front row seat
And get all involved in the play
Or it can sit way back in the theater
And put some distance
Between itself and the drama.

As there is no escape from existence,
For it is always in front of our nose,
One may as well enjoy it,
For this is life,
Although, at times,
One may stand back,
Disengaging and detaching,
For whatever reason,
Just letting the scenes or thoughts go by,
Like a parade, with much less involvement.

The above is another method of meditation,
Other than removing all thoughts,
Although that, too, will work, later, but first,
Let the events or thoughts enter the stage,
Pass by, and exit, all from a sobering distance.

When we watch a good movie or read a good book
We purposely let ourselves get deeply into it
So as to have the full enjoyment of it,
The same as in living a good life.

However, if the movie contains too much terror
We might then distance ourselves from it,
Noting that it only involves actors,
Props, light on a screen,
And that there are cameramen unseen
Off to the side, and a microphone hanging down.

DIPLOMACY

The best diplomat is one
Who can tell someone
To go to a place really warm
Such that they feel that
They will even enjoy the trip.

BRAIN STORM

The brain does all of the analysis
Leading to thoughts, ideas, actions,
And so forth, via our memories,
Associations, senses, and learning—
Which the brain contains and maintains.

The brain is also known
As the mind or the will,
But it is much of the same.

The brain may be complex, in parts,
Or it may often be a lot of simpleton areas
Ever competing for attention.

Whatever result surfaces
Is observed by consciousness,
Another mode of the brain
Which perceives the rest of the brain—
Kind of like another sense—the 6th.

Brain results are either accepted
By the overall brain
Or are fed back for further analysis:

"OK, I'll do it."
Or
"Let's find a better furniture arrangement."
Or
"What will I do today?"

Sooner or later the brain gets back to us
On what to do if we ask it to plan something;
Although, when we ask the brain to ponder
And solve the Theory of Everything
And then get back to us, it never does!

RANDOMLY DETERMINED

If what wills the will is not done randomly,
Then determined it must then be;
But, what of ties where there is no rather?
Well, one tie may wear as good as another.

IMAGINATION

We can imagine pink cows, purple cows,
Flying cows, cows with heads on either end,
But, we mostly restrict ourselves
To the cows we've seen or heard of.

We've actually seen cows
With two heads on the front,
But no flying ones.

One time someone proposed a black swan,
Which was really not so unreasonable,
But this was before every nook and cranny
Of the world had been explored.

A person suggested that every time
He saw yet another white swan
That this surely added more proof
That all swans were white.

Well, I wouldn't go that far,
For it only added a tiny bit of tendency;
Only if you surveyed every square inch
Would it carry some real weight.

Then, indeed a black swan was discovered,
Way back when a certain area of the world
Was just beginning to be gotten into: Australia.

THE SPLIT

The right brain is the mystic,
While the left brain is the scientist.

A right brain alone
Could only see the unfocused whole,
While the focusing left brain alone
Wouldn't have an overall view
From which to select any specifics.

However, it seems that it's hard to be
A scientist and a mystic
At the exact same time,
Since the left/right brain's methods
Are so opposite.

INANIMATED TO ANIMATED

Perhaps the knowledge of movement
Makes for animated life,
But, when did this happen?

How does one draw a clear line
Between organization and not?

When, even, does the night turn to day?

The most interesting and potent things,
From the evolution of the universe to life,
Exist at the blurred boundary
Between order and chaos...

Life perhaps emerging in tide pools—
The shifting edge between land and sea.

It is all of the fuzzy realm
In which and where things
Have to be orderly enough to take form,
But not so much frozen that they cannot change.

COLORS AND FORMS

While the primary colors are seemingly
Arbitrarily chosen, this does not matter,
For they would ever help us distinguish reality.

The capability varies from species to species,
And perhaps there is even
A slight variance among humans,
Although probably not as much
As with the taste of food.

My cat only sees in black and white (gray).

Its still interesting to know how
The entire visual system of the brain works,
Finding edges and shapes of objects, and all that,
For vision seems to pretty much dominate perception.

For color interpretation,
Three types of proteins in the eye cones
Rotate according to the amount
Of their primary color received.

NEUROLOGY

Brain cells (neurons),
Of which there are a hundred billion,
Each have thousands of connections,
Their "firing" dependent on their inputs.

Electricity carries the "message"
Through the length of the cell
To the gap (the synapse),
Where the message turns to chemical
(neurotransmitters)
To take it on to another neuron
Wherein it becomes electrical again,
And so on.

Your brain neurons
Have been arranging
Their connections all your life.

It is what you have become,
Molded by your experience and learning.

You are a bio-electrical-chemical being.

AETHER

In general,
About electromagnetic waves...
Let's think of space anywhere:

A changing electric field
Produces a magnetic field
And so then that produces
A further electric field
And so on, this being
A self-renewing disturbance;
That is, what we know as visible light,
X-rays, infrared, and so much more
That are of the electromagnetic spectrum.

Thus, no aether is required
As a medium for electromagnetic waves
To propagate from here to there.

So what about these fields
And those of the 'vacuum'?

Well, they are kind of like an aether.

FIRE AND ICE

I am not a 'the end is near' kind of guy,
But it's fun to write end-of-the-world stories
And thus work in some drama, science,
And the histories of extinctions
To demonstrate the larger scheme of things.

When all of Florida went under,
Being that it was already sea level,
The world woke up, but it was much too late.

A year later, every coastal city in the world
Had been inundated, for water always finds its way in.

Northern Canada, along with its eastern
And western sections, was no more;
Only the now more temperate interior remained dry,
It filling up with more and more people
And tropical birds by the day.

Mikal had moved to the cabin,
For Lake Ontario was acting up;
A small portion of Siberia was usable, too,
For now, and an 'iffy' part of Argentina.

Alaska was gone to its mountains,
As were Central and Latin America, and Mexico;
The heat was melting the ice caps.

The equatorial regions had become unbearable,
The temperatures now reaching 135 degrees F;
Most of these people really had no place to go.

The pace of the disaster was startling,
Exceeding even the grimmest of predictions.

Australia's population, being mostly coastal,
Had retreated to the thin edge between
Its useless interior and the deep blue sea.

Graybeard was holding on,
Being one who was acclimated to extremes,
As well as to living off the land.

He'd climbed a sturdy tree
When the tidal wave swept inland,
Perching there for a day until the waters had receded
Enough for him to slog his way home.

His tiny car with its 2-horsepower engine
Had been washed away,
But he still had the big one at home;
His ranch was soggy but otherwise secure.

The Great Inland Sea reappeared,
Then evaporated, and then returned;
Meteorology had become a fruitless science;
Summer was now year-round.
He readied his camping gear just in case.

The Yukon was a mess, too;
Yet, LabelWench, as well, was a survivor;
The dogs had been leashed, the sled provisioned.

Cars could not brave the mud and the slush,
So she was off on the ultimate Yukon Quest
With her sled and dogs. Whitehorse was dying.

ProfPat was fine where he was,
Although the Great Lakes were enlarging,
Even with their locks having been closed,
For the water runneth over.
All this from a Bindu dot, he thought.

Resistance was futile,
For water was slippery stuff,
Its tiny hydrogen atoms
Easily rolling around the oxygen.

Various schemes and solutions
Had been endlessly debated,
Most of them being
As tough on mankind as the heat,
The rest of them unattainable;
It was all happening much too quickly.

The estimates had been way off—
Nature was now a dragon and she had roared
And was spewing her fiery breath
Upon her own fragile planet.

Some tried to live in basements,
But the heat found them;
Some dug underground, but, the floods came.

Many near extinctions had happened before,
One being only 74,000 years ago, when,
At Toba, in northern Sumatra,
A supervolcano erupted.

Six years of volcanic winter followed this eruption,
Bringing pre-humans to the very edge of elimination;
There were but a few thousand of them left around,
Since very little light could reach the dusty ground.

It took twenty thousand years for them to recompose
From the caldron of fire
That had almost brewed humanity away.
It was from this handful of hardy souls
That we modern humans arose.

Back to the heat...
Due to the intense global warming and ice melting
That was about to reach a runaway exponential point,
The weather patterns had been greatly altered
And were bringing numerous and severe storms
All over the globe, causing much interior flooding,
Plus, wiping out most of the crops.

The rate of destruction
Was becoming astronomical.
No one could keep up with it;
It was everyone for themselves.

2012 had long since passed without incident,
And now it was 2021, an anagram of 2012.

Billions had perished in a matter of a few weeks
And there was no government infrastructure left,
Save for the covert Ninja Empire;
However, humans were hardly a match
For nature gone wild;
It could and would only get worse unto the end.

(The now worldwide Ninja Empire
Began as a reaction to the Conspiracy
Such at that of the 'Nowhere Man' TV series;
They also recruited scientists from the Lists and Forums
And discovered the TOE, it having to do with the
DNA of the universe broadcasting through the CMBR.

At first they were chased all over by the CIA,
KGB, and those desiring the TOE,
But, they eventually allied with them
And much broadened their efforts—
To include terrorism and more.)

Back in 1960, Bob Christiansen had looked around
Everywhere at the Yellowstone National Park grounds
For its volcanic caldera, but had found it nowhere.

By some coincidence, NASA had photos
From a recently tested high altitude camera;
Astounded, Bob learned why he'd failed to spot the caldera.
It was virtually the entire park—2.2 million acres of area!

Yellowstone must have blown up with a violent misery
Far beyond anything known throughout our history;
The crater was forty miles across.

There was also a huge gap in the mountain chain.
The cataclysm had been even beyond the scale
Of what imaginations could have dreamt up—
It had thousands of times more monstrous molten fire
Than Mount St. Helens;
Krakatau was but a firecracker in comparison.

Yellowstone's eruptions average one really massive blow
Every 600,000 years, the last one being 630,000 years ago;
So, it was long overdue and perhaps pending,
But, still, that could be thousands of years off.

The Ninja leaders and their teams flew into Niihau
On their last tanks of gas, there to take sanctuary
And a vote for a desperate but uncertain plan.
The landing strip now had quite a tricky slope to it.

Many near extinctions had occurred before,
But none in modern times.
However, death seemed to be a way of life
Throughout history.

Of all extinctions, the Permian was the largest,
Happening 245 million years ago,
When 95% of species perished,
Suddenly disappearing from the fossil recording;
Life had almost come to a total obliterationing.

That we are even here was due to the dinosaurs
And 90% of all the species being wiped out.
A small and nervous shrew-like creature looked out—
The dinosaurs, the forever Kings of the Earth, were gone.

The 'shrews' then attached to a favorable evolutionary line; Every
single one of our forbears on both sides survived,
They being attractive enough to locate a loving mate
With whom their love to celebrate.

In the late 1700's, Cuvier could take heaps of bones
And whip them into shapely forms not in the stones.

After naming the fossil elephant the mastodon,
He put forward for the first time a theory on extinction.

He said that from time to time there were global catastrophes
In which some groups of creatures "became history".
This raised uncomfortable implications at the time:
Why would God create & destroy without reason or rhyme?

This suggested an unaccountable casualness by
Someone unseeing and greatly troubled the belief in
The Great Chain of Being,
Which held that the world
Was carefully ordered for us
And that every living thing
Thus had a place and purpose.

Meanwhile, William Smith had noted a correlation in fossils
In rocks to find the relative rock ages that were possible:
At every change in rock strata, certain fossils vanished,
While in others they carried on into subsequent levels.

Now it was seen that Nature (God) had wiped out creatures
Extinct not only occasionally but repeatedly—
Which made one think Him not only careless
But having an outright hostile distinction,
More than the Biblical Noachian deluge extinction.

Of course, now we see nature more for what it is.

Back to the future...

It no longer mattered that humans
Were on a one way trip from the quantum fluke,
That maximal disorder within old Planck's nook,
To the escalation caused by the Dark Matter,
The Universe heading toward the oblivion
Of its sparse and accelerated expansion,
All that we ever loved and knew going to extinction;

For, the world was ending now,
Or at least it was the beginning of the end,
For the temperatures were increasing daily.

Mountainous Niihau was somewhat unaffected,
As were other such high island regions,
Although they were few and far between,
For which the trade winds still brought some moderation;
Yet, the subtropical and near tropical areas
Were now all very much overly tropical.

Al Gore had been the last one
Into the Washington bunkers,
Having steadfastly worked up to the end.

Some of the remaining 'nations' of the world
Noted that they had indeed
Squandered the good things of life,
Squabbling over differences in culture.

Many had all been looking for truth
In the wrong direction,
Which was back into the past
Instead of to the future—
They had fiddled,
And now Rome was burning.

On Niihau now, Fredrick walked the last mile.
The world's roulette wheel was rolling double zero
On what might be the last perfect day on Earth,
And a hurricane was headed in his direction.

Tsunamis washed deeper and deeper inland;
Most of the internet no longer functioned;
ToeQuest and AVOID were gone,
Their archives destroyed.

'Nobody' had returned,
His efforts at changing the future
Having been unsuccessful.

Robert gave up his dreams of the TOE—
The TOE center on Oahu had closed.
Graham finally levitated up into the air,
His new refuge above the clouds.

Trish—#1 East—energized her way
To the Forbidden Island,
Noting its re-creation of the Forbidden City;
Another top dragon, Old Rascal—the General—
#1 West—awaited them all.

A rainbow appeared as the spirit
Of old GrandMaster West
Flying through, past them,
As the scent on the breeze
And the courage to act.

The votes had been counted.
Rascal had now engaged the last working silo;

Trish keyed in the codes and nodded to him,
Saying "Would you like to do the honors?"

"Yes, for unto me falls life's last duty.
Nature will now have to contend with itself;
It is fire versus ice, dragon against dragon;
However, this could well be the end for all of us."

"Or not, if it works," she answered.
"The years will tell."

Rascal pressed the button:
Nuclear missiles were being fired,
Deep into the heart of the Yellowstone caldera.

Five surrounding states and two Canadian provinces
Would be completely destroyed; yet, winter would come,
Although probably bringing its own array of problems.

Drive my dead thoughts over the universe,
Like wither'd leaves, to quicken a new birth;
And, by the incantation of this verse,
Scatter, as from an unextinguish'd hearth
Ashes and sparks, my words among mankind!

Be through my lips to unawaken'd earth
The trumpet of a prophecy! O Wind,
If Winter comes, can Spring be far behind?

Poem by <u>Percy Bysshe Shelley</u>
English poet (1792 - 1822)

FOUND KNOWING

Well, we knew that a little learning was dangerous,
And now thinking, too, for minds can be read through,
Well, my thoughts are sane, so it's OK to read a few,
And the rest of them I'll put into poems profoundest.

HOYLE'S RETURN

Old Hoyle's ghost walked out the gate,
Then returned to his living steady state.
He gamed with the cousin Hoyle's lot,
Throwing the die away, upping the pot.

NEW LAWS

These are the new laws and formulas
That reflect my failed attempt
To convert to Christianity...

Force of Gravity
=
SomeFactor*[(Mass1*Mass2)/Radius squared]

(law of gravitation)

(+ motion added)

This law, discovered by Newton under a tree
That is still bearing falling fruit,
Along with some other law of his or someone's
About a body in motion staying in motion
And a body at rest staying at rest,
Each of their own accord,
Was thought to cause planets
To circle a star (sun),
Although they did so as an ellipse,
For which there is also
Some kind of wonderful formula,
And so it is that if a planet had no sun
Then it would just go in a line, not ellipsing,
And, if it had a sun and it were at rest
It would plunge straight into the heart of its sun;

But, since being in motion and having a sun
The planet then takes the in-between path,
As it literally falls around the sun,
But, who cares,
For none of this is true anymore.

In actuality,
God guides the planets safely around the sun
Through their orbits, so...

GM (Gravity&Motion) = GGH (God's Guiding Hand)

Many textbooks will have to be changed
To show this new truth.

Perhaps we could refer to it as

Isaac: Revelations II.

H2O
(Chemical formula for water)

This is the most often used formula.

It is thought that having
Such a small molecule of hydrogen
Attached to such a large molecule as oxygen
Causes the sliding around
That makes water molecules so very slippery.

However, it is really that God micromanages
Every single atom
And also even what is inside of them.

Besides, God's son showed
That water could be transmuted, so...

H2O=>WINE

Jesus thought of opening a brewery,
But then had another calling.

E=MCC
(Einstein's conversion of mass to energy)

A little mass makes for a gigantic energy
Such as that of a nuclear bomb,
But, really, although bombs work,
Pay no attention to Einstein,
For, actually, it is
That God's very powerful Energy
Makes our energy.

There is no need to worry about
Where God's tremendous energy comes from,
For God is actually a semi-unified twin-genii
Split into Good and Evil,
Positive and negative,
And so He is zero on balance
At the end of the day.
The same for good and bad angels.

Zero=Good+Evil to the infinite power.

**A squared + B squared = C squared
(Pythagorean theorem for a right triangle)**

No one cares about this any more,
Nor even any Geometry,
For God can make a square circle.

From now on this formula
Refers only to the Holy Trinity.

God = 3 = DoAnything

**Very Complicated Formula
(radioactive decay of uranium)**

This has been replaced by God's power of alchemy,
For that's what it really was anyway.

**Some kind of -b+-
some square root thing over 2a, etc.
(solution of quadratic)**

By memorizing this great formula
That I have now totally forgotten,
I was able to do very well in advanced algebra.

This formula will be of no use
To us in advanced theology,
Which, as in all of my seminary courses,
Will just say that 'God did it'.

The Shortest Course on 'Whodunit' Theory

I took a really short course called GOD 101:
It only lasted about 3 minutes,
And there was no continuation of 102, 103, etc.

The instruction consisted of but one statement:
'God did it.'

Note that this statement
Not only puts the answer before the inquiry
But that it also halts all inquiry;
Thus, the case is closed before it can open.

Really, God is only a theory,
But I didn't let on,
Answering 'God did it'
On the 1-question test of
'What is the answer to anything?',
Thereby passing the test,
The course and the college.

The metronome graduates
Never said 'God is a theory',
But ever produced a regulated aural pulse
Of a steady tempo, in saying:
'God did it.'

The mind makes funny shortcuts
For what it cannot know;
But, I am not one of those minds.

A—T, C—G
(four building blocks of DNA with their matchings)

It has been found
That these letters spell 'God' in old Hebrew,
So, that is the story of that one.

A bunch of functions that I hate
(fundamental theorem of calculus)

I never liked figuring out all
Of these unholy moving things,
But now I am unmoved by all things holy.

$V=RI$
(Ohm's law)

Ohm has been found to be ooommmm,
The focus of meditation—
On God or of a sleeping dreaming Brahman.

$Pm = Po(some\ crap)$
(compound interest)

An accountant will redo this one,
But it has to do with
Accumulating treasure in Heaven.

TOE

This equals the Ground-of-Determination.
Note that the acronym for this is GOD.

SOUL
(invisible appendage replacing the brain)

It stands for:
Spirit-Of-Unconditional-Love

Quark

A marking was found on this material:

Holy stuff:
Made by God

Evolution

Explained by anagrams:

Outlive On
Olive Unto
Ovule Into
Vile No Out
Vile On Out
Live No Out
Ovule It On
Love I Unto
Love In Out

=

No words can describe this symbol,
For it is what it is.

THE HISTORY OF LIFE

DNA remembers evolution and natural selection,
Playing back billions of years
During the 9 months required to build a human.

TO THE SIMPLE

Going back down the scale,
Though the atom and on into the quark—
Or string or whatever—to the simplest point—
Is certainly not where one would expect
To find the most ultimate complex composite
Of all time and size to the infinite power.

Well whatever took 14 billions years
To form man was near brainless,
For what could be so brainlessly slow
Other than evolution via natural selection?

My own invented God would be all loving,
But that would only be wishful thinking.

In general, for what has never shown,
Never been known, never flown,
And never sown, etc.,
The human concept of God, is,
By some wild coincidence,
A Higher Person of human qualities
Of emotions, tantrums, strictness,
And of much fatherly discipline.

PROGRESS

Astrology gave way to astronomy,
Alchemy to chemistry,
And religion to philosophy and science.

Discovering truth provides freedom
From the shackles of myth;
It is not doom,
As the notion of Pandora's box
Deceptively paints it.

The box of truth opens by itself,
No matter how much
One might try to put a lid on it.

ALL AND EVERYTHING

Perhaps when we figure out
The quantum realm's manufacture
Of particles out of superposition
Of its potential and possibility,
Which would be the most major find ever,
Then we can go on to the potential
Of what made the All—
The energy/matter/quantum realm itself
Coming into being as the origin of the universe,
This probably making the quantum realm workings
Then seem as child play in comparison.

Of course,
This is all "before" where Jimbo started
With his Fluid Energy Theory,
He beginning with energy
And not worrying too much about
Where that came from but to say
That it was around forever.

I could see that if it was
Truly liquid-like energy
Then its exact nature would not
Had have to be predefined
Exactly, within some range,
Although really huge stuff
Might not have worked at all;

But, what about the amount of it
And where it was and why was it there
And probably many other things
That could never have been
Specified in the first place,
Since there was no first place,
The stuff having been around forever.

That, too, would mean
That some kind of eternity
Had already happened,
But, eternities never complete,
Nor do infinities, so...?...

I guess that's why we need a beginning
That was neither of nothing nor real stuff,
Such as the in-between state
That we might call "potential",
It being some kind of possibility structure
That needs no causes behind it,

For what would those be
Except for more possibility—
And we already have that
And so we don't need more of it.

This, too, gets around the problem
Of the infinite regress of stuff
Always being made of lesser stuff.

The big clue of an absolute Nothing
Not having succeeded in preventing something
Pretty much proves that
There HAD TO BE something
And that this [potential]
Could not be stopped,
Although, of course, had there been
A total absolute Nothing
Then certainly nothing
Would ever have become of it.

MYTHOLOGY

The lion lies as the way the sphinx thinks.
Aion is the evolutionary ancestor?
Our 'knowing' is the Greatest that never was,
Since, to say, it just does what nature does.

HER SODIUM LEVEL PLUNGED

Her sodium fell, then she took a fall,
Nearly dying, evolving beyond it all...
In a NDE while she was not much alive—
But she found that evolution's aim was to survive.

So, could she have lived then,
With the low sodium?
Well, they always said we could,
Some of those those unknowing medicals,
Saying that a low salt diet is good.

She might have to study
The evolution of the species,
To know why and whence we came
Out of those salty seas.

WHAT, ME WORRY?

As for worries,
Most of them never happen;
So don't worry.

If they still happen sometimes,
No matter how prepared you were,
Then it's still good that you didn't worry,
For they were unavoidable.

Anything wrong with this logic?

As for getting angry at something
Like spilling milk, there's no need,
For you still have to clean up the milk—
And its easier if you are not mad.

As for answers, they are really easy,
Being "yes", "no", and "maybe";
It's the questions that are hard.

Note that saying "maybe" to a child
Or to a romantic partner means "yes".

WHEN GENDER IS UNKNOWN OR IRRELEVANT

When gender is unknown,
Some pronouns can't be said,
So, for he or she, use "e",

As for him or her, it's "erm",
And for his or hers, use "eir".

As for a singular you, that is it;
For the plural, use "you-all".

HUMAN BEING DEFINITION:

Ignorance in action of much ado about nothing.

ESSENCE IS EXISTENCE

There never was, nor will be; but just 'now'—
All things, interacting—planned know-how,
For mind "matters", matter ever "minds"; so,
The Universe self-adjusts—like the Tao.

The Exacting Parameters

The precise tuning of the Universe
Was performed recursively, in reverse,
As it looped back and fed into itself—
For it exists all at once—every verse.
(potentially)

Universal Creativity

Brains, the greatest that matter ever wrought,
With the mind, formed the Cosmos that was sought.
Observation's interaction became reality—
(via consciousness)
The necessary constants being self-taught.

Unreal-ized Power

Mind reaches out to see what's possible
And what's not, like particles forming
In the quantum world, but, better than that—
It makes the potential possible.

Math Before Existence?

Matter is merely mathematical:
Wave-function possibilities fall
Out of eternity's equations;
Someday, mind will grow to encompass all.

The Future Past

The Universe could only be created
In its own future—through observation
Of 'mind' granting substance to formula—
Even from an eventual single, eternal Mind,
Bringing all into existence,
Either as is, or from the small.

WITHIN AND WITHOUT

Life takes me everywhere,
Even deep in here,
As I have shown
As well you've known.

These human traits are dear
From stars that brought us here.

So, this mammal knows
The depths of "soul",
Translating it as what flows
Into poems that rock and roll.

CHARMS

A new kind of microscope
That works via gravitational waves
Has revealed the actual interior
Of a quark for the first time.

The charming beauty of the ultimate truth
Is that ladies are in charge of the universe!

REVOLVING

Hard to put a "spin"
On this world we are in
To say of what neither twirls
Nor whirls or swirls.

With nothing around it,
We can't spin it,
But if we did it,
We wouldn't notice it.

Maybe a basketball caught
In a court of mostly naught,
With none around,
I might hold still,
But then I might catch a cold chill.

**INTRODUCTION
OF
AUSTIN[TORN]
TO
THE TOEQUEST FORUM**

What to Say

1.
Where are you from?

I am from stardust,
And of protons before that,
And reside in Poughquag, NY,
On a mountain top
Near the appalachian Trail.

I have not lived here all my life,
As my life is not over yet.

2.
How did you locate the web site?

Googling and oogling
The "Theory of Everything".

3.
What got you interested in the Theory of Everything?

I wondered what the heck I was doing here,
Thrust into life on Earth in a galaxy
In the middle of nowhere.

I would say something like
Whatever Brought Me Here
Will Have To Take Me Home,
But Rumi already said something like that.

4.
Do you have a unique perspective?

Only to connect ideas into other "new" ideas
To lubricate the Toes so they won't go dry,
Plus, to add some humor to keep things light
And also to post a lot to keep things lively.

5.
Where did you go to school and what did you study?

I attended the University of Illinois and Hawaii,
Studying computer science and student bodies.

6.
What areas are you most familiar with or interested in?

Reading, writing prose and poetry, physics,
Nature, neurology, life, tennis, and philosophy,
But I don't know much about history.

7.
Have you written any books or publications?

I put a bunch out, On Amazon,
As createspace.com lets you do that for free.

8.
What past influences have shaped your opinion
Of our world and its greater reality?

This would be family, culture, education,
Various leaders, scientists, religion,
Personal experience,
Omar Khayyam and the Big Bang.

What not to Say

1.
Please don't tell us that you already know
The Theory of Everything.

You should try and get involved
With the forum discussions
And become a welcomed
And respected member
Of the community before you drop this on us.

The TOE is that we can't KNOW,
Plus, that all was uncaused. QED

2.
Avoid starting a science-related topic
With very little or no introduction of yourself.
We have many other forums for science topics.
Sometimes a topic discussion ensues from an introduction,
But that should not be the purpose
Of your introductory post.

OK, but too late.

:

3.
Please avoid any preaching,
Political discussions, spam, etc.

Luckily that I am not religious, political,
Or selling Avon, so this part went OK,
But I did know Obama in both Hawaii and Illinois;
However, I can't talk about that.

Let the QUEST begin.

— Austin

INFLATION [OF OURSELVES]

The last few centuries has seen our species
Gain a perspective on the universe
Unprecedented in history,
And this has expanded
On some "inflationary" sense
Since early-mid 20th century.

Along with this has been
The decentralization
Of humanity in the universe,
To the point we might be seen
As little more than cosmic debris.

AFTER DEATH

When everything expires of that fate,
I'll still try to drink after the expiration date.

THE POTENTIAL OF
THE POSSIBLE-PROBABLE REALIZED

Things with a real definition could not
Have been around forever,
As they would not have had any definition
In the first place (for that there never was),
Plus, this already completed eternity
Of time—the movement of the definite—
Unto now from that endless stretch
Forever back to no maternity—
Could not have been completed,
As infinities and forevers
Can never ever can be.

What could there be 'that' still could have been
Around forever, needing nothing prior
(For that would only be itself)?

It would have to have been indefinite
Being only of itself.

It could not be a thing,
Since it needs to be
Both formless and timeless.

It cannot be a system of mind,
For that is dependent on its parts
Of energy/mass—
Those being more fundamental things.

It could not be Nothing,
For nothing would ever become of that
Which is not even a "that".

It could not be of physical laws,
For there were none before it to form them.

You see, all was wide open "then"—
And so it was Possibility that was forever.

So, there's a non-formed
Fundamental Potential Of Possibility,
This being all things (universes) in superposition
That are happening and evolving all at once,
Even rather instantly,
As all is timeless in its formlessness.

One of these possible potential paths
Generated our workable universe
In which creatures went on
To achieve consciousness.

This consciousness of those
In the projected 'future'
Affected all of the 'past'
By bringing it forth into
The realm of the actual—
A seeming paradox of
Some kind of feedback loop.

The future looped back,
Like the water that
Evaporates from the ground,
Going back up to the clouds,
Then raining back to the ground.

So, it could be that all universes evolved
But in potential form,
Until some great mind,
Or any mind, either past,
Now, or in the future,
Brought forth the real into the actual
From its potential state of possibility
By virtue of it having a conscious observation,
As kind of like in the quantum realm
When particles are summoned
Or at least hastened into being,
And also because the
Next dimension beyond ours
Would naturally be of the nature
Of all possible occurrences happening in potential.

We live in one of these
Shimmering paths of everything,
Our own pot of gold being within its rainbow.

TRASH COMPACTOR

Insults are not at all an acceptable rapture;
So please put them in the trash compactor.
If prospective members see ever more wrongs,
Then they'll all be off to sing different songs.

IS KNOWLEDGE OF ESSENCE EVEN IMPORTANT?

Even though we seem
To 'know' our essence of matter,
Is this part really that important?

Yes, in a way, perhaps,
For some sense about
The human condition;

But, after that, "no"—
For, during much or all of our living,
We are thrust into this life
And so we must deal with it,
And thus live it first and foremost,
That needing our primary attention,
Making it "precede", in importance,
Any exact knowing of the essence.

Is the 'possible' still arising
And forming reality instant by instant?

Perhaps, but it doesn't matter,
For that is just the mechanism;
The more important content
Is of the message of our reality,
Being that which we receive
As life from what is beneath it.

Is matter a holy thing?

No, it just is what it is.

Do we go away
And then return here, recycled?

Yes, but you would not be
You again as you were.

What should we really do
About all this existence?

Well, live it;
Don't just merely exist!

How did we figure
This 'Possibility' angle out?

We were merely redirected
By some road blocks
Of commonly sought avenues
Into a dark alley where the truth
Was just waiting to show us its light.

THE SELF-LESS ONES

Mystics and Mysters ravel the TOE threads
With those answers that can never be read;
So many words unreal—now they are spent;
Thence their descent, hence toward oblivion went...
Into the bliss of nothingness that dreams meant.

HELPING

There are some remoter shores of human souls
To which I'm helping restore life and spirit;
My prior life was but a preparation
For these trials and pains that love must give.

Deep they fall, into valleys of despair,
Although for those the mountains wouldn't be there;
But, it is difficult to see the tip
When you are at the bottom of the pit.

They have many feelings, many not true,
Taking them beyond, into the blue
Of depression; but, abandonments
Were of those not able—no reflection on you.

CONSCIOUSNESS IS THE SOUL?

We know that consciousness is but of the brain
Because we can introduce molecules into brain areas,
By anesthesia, and, thus,
Turn off consciousness completely;
So, consciousness has no independent existence.

STRANGE THEORIES

Our energetic anatomy is exactly like the Earth's,
With chakras, meridians, axiatonal lines, and DNA.

Our 13th chakra is actually one we share with the Earth
And exists at the Earth's core;
So, if a cataclysm happens with the Earth
And it pokes a hole in the Earth's grids,
Then what happens is that "hole"
Is poked into every person's DNA on the planet.

Well, Earth has no DNA but that of the plants on it.
DNA holes don't come from poking the Earth's core
Or from cataclysms on Earth;
Actually, cataclysms helped natural selection
Move along more rapidly.

Someone has a hole in their head,
But that is beside the point
[Which they also have].

And this has been one of the main influences
And problems with our DNA throughout history,
As there have been many cataclysms.

The Earth distortions in history have actually
Created a planetary DNA distortion
That has made every living organism's DNA on the planet
To falsely appear as Base-4, with only 4 chemicals.

Humans are supposed to have
12 chemical nucleotide bases;
This would allow us to have
144 physical chemical chromosomes as well,
Whereas now we only have 46.

We have 23.
Two of them merged,
As we used to have 24, like the chimps;
This allowed us to go our own way,
For then there was no more monkeying around.

There is no "supposed to have 12",
And the Earth is only doing just planets do—
In its happenings of weather and ground.

Also the original Angelic Human DNA Template

Which most people on the planet have,
Called the Diamond Sun DNA Template,
Is 12 strands,
Allowing for 12 dimensions of consciousness,
And is built for transmutation
From carbon to silica based body,
And eventually to pre-matter liquid light.

There is no angelic template nor angels doing this,
Nor other assorted invisibles claimed to be involved in DNA.
DNA is chemical. The only silica is in computer chips.

This body is not meant to "die".
Death is because of the DNA distortions
In ancient history (25,500 B.C.)
That have blocked people from bringing light into their field
And naturally evolving through DNA activation.

Most people on the planet only have 3 strands active,
Which only allows 3 dimensions of consciousness
and thus they are stuck in 3D.

This is a category error;
The three dimensions have nothing to do
With the 3 strands of DNA, whatever they are.
I thought DNA had two strands.

25,000 years is but an evolutionary instant.

If we were not meant to die,
Then I guess Somebody goofed, for die we do.

We are at a very important point in history right now
Because we have the ability to regenerate
Our original organic imprint for health.

And we don't need science and fancy equipment to do this.
We are just required to learn how
To use what we came in with—
Our mind-body-spirit system to direct our mind
To alter the scalar waves by which our DNA is composed.

Made up boloney.

We are in the middle of an ascension cycle,
Which is literally a time continuum shift.
From now through 2012;

The planet is going through this time continuum shift,
Which only happens once every 25,556 years
(called a Euiago cycle).

What is happening is that the particles that make up
The Earth's auric field are speeding up
In pulsation rhythm to prepare for this shift
Into ascension from dimension 3 to dimension 4.

Since the particles that make up the Earth's auric field
Are speeding up in their pulsation rate,
And we exist as part of the Earth's auric field,
The particles that make up our OWN auric fields
Must also speed up.

This is happening now through 2012.
This means that one must have at least 4 strands
Of DNA active to make the shift into dimension 4.

Nothing of this nature will be happening
During the next few years.
(Or I guess I will be stranded.)

And this is why DNA activation is so important now.
If you have energetic blockages –
Auric attachments, karmic imprints,
DNA distortions, or unnatural energetic seals,
Then you will not be able to speed up the particles
And accrete the frequencies necessary
To make this time continuum shift

And the higher frequencies coming into your field
Will speed up the body deterioration process
And many result in physical problems or discomfort.
Many people are noticing these "symptoms" now
And this is why
They are also noticing how time is speeding up.

Anyone out there having extra aches and pains now?
Time seems to go by fast on the internet"
My energetic karma seal
Is hopping along like a kangaroo.

Our evolution is directly and intimately tied to the Earth's.
And the actual purpose of the Angelic Human
Is to assist the Earth with what is happening now.
This IS "why we are here".

Because by raising our own frequency
And bringing more light into our fields by DNA activation,
When enough people do this and it meets "critical mass",
This will actually reset the distortions in the Earth's grids
So that we can all ascend together in 2012.

All masters that have come, are here,
And will know about that DNA activation
And spiritual evolution does not happen without it.

In 2012 we can revisit this thread,
But by then we may all be dead.

Summary:

About the winged lobes
And the angelic wars
And the 12-dimensional DNA...

I was not offended at all;
That is not at all my nature.
On the contrary,
I received a great deal of amusement
From the reading
Of the never ending tendency
Of human nature towards superstition,
Not to mention them ignoring science
And/or getting it all wrong.

These types of posts are always most welcome.

I'm not sure why the angels went to war
Against God if He is so great,
But... please don't tell me—

I am not drunk now, nor smoking pot,
Nor mentally unbalanced,
Nor watching sick TV sit-coms,
But, I'm not sure if it's useful to follow
The angelic twists and turns
While one is unstoned,
For the worms beneath the rocks
May start to bug me.

THE QUICK AND THE DEAD—
ON THE LAST AUSTRALIAN MUNDI

Graybeard (Greg) headed for
The waterless Mundi regions,
Where the winds and the sands
Sculpted and streaked the rocks,
And where the Knights Templar
Of the armor plates
Would be at a disadvantage...

There he waited and looked up
At the sharp white stars.

Soon his pursuers would arrive,
For he had let it out
That this was his destination;
However, all was never as it seemed.

On the Last Mundi, or was it Tuesday,
Greg (Graybeard's alter ego) was walking
Along the windswept plain of the Mundi
On his way back to his camp at the large rock,
Returning from a hike in the mountains.

It had been a good day with nature;
He already felt somewhat primeval.

It was almost dusk
And so the stars of home
Would soon shine above and beyond.

It was good to get away to see and learn
What more that this life was all about.

What's that!

A mad rabid dog ran out from the shadows,
Heading crazily but swiftly toward Graybeard...

How did he know all this?
Of what is a human made?

Would an acute fear response
Give him a good shot at staying alive?
Should he confront or avoid?

In this case he would have to
Try to avoid by flight
And then perhaps confront by fight,

Which is really more like a freeze,
There being flight, fight, or fright.

We are actually hardwired to flee first,
But, if overtaken, we must defend,
Although a trancelike passive state
Of being filled with fear is also possible.

Within seconds, Graybeard was primed;
His pupils were dilated
And his respiration had sped up.

He stopped producing saliva
As sweat poured out all over;
His blood rushed away from his stomach
To soak his brain and muscles
In nutrients and oxygen,
Energizing him for what lay ahead.

He froze, watched, and listened
But only for a second.

Light waves flashing off of the dog's teeth
Had passed through his eyes
That could now see all the better.

Electric signals had entered his brain
The visual information
Routed to the opposite sides,
This depth perception helping him
To better locate and keep track
Of the oncoming and insane assailant.

The sound waves of
The dog's growling and barking
Crashed against the tympanic membranes
Of his ears and were on into his brain
As sounds to be processed.

Yes, all this, as well, in a second or two.

Within milliseconds,
Neurotransmitters chemically
Ferried electric signals
From one neuron to the next,
Spreading the latest news of the dog
To a quick response unit
Born of ancient times.

The sensory information had funneled
Deeper into the brain for further analysis.

Graybeard's vast network of neurons
Lit up like a Christmas tree.
The ultimate decision would be made
By the amygdala—the "fear" center.

Would there just be cause
For a temporary alertness
Or should there be
A full-fledged fear response?

The dog was going wild;
There were no trees to climb;
There was little chance for escape, but....

The amygdala sent a siren sounding
Through Graybeard's brain,
Having cued the locus ceruleus
In the brain stem to release gobs
Of the neurotransmitter nor-epinephrine.

Twin brain structures
At the bottom of the head,
The cerebellum,
Considered various attempts
For escape or of self-defense.

All of Graybeard's ancestors
Had now arisen to heed the threat.

The brain stem had sent
An all-points bulletin,
This constricting his blood vessels
And inhibiting all ordinary
Parasympathetic nervous activity.

His throat tightened
In case a scream was necessary;
His body was preparing for the worst.

This real danger was
What his formerly safe life
Had come down to.

To stare death in the face
Was now to live twice.

The dog was fifty yards away
And bearing down upon him,
Its nature having gone wilder than wild.

The spinal cord had aided the cascade
Of the acute fear response
To all the corners of Graybeard's body,
Activating the peripheral nervous system
Of his arms and legs, among other senses,
To attend to stimuli
Of the new and dangerous environment.

Greg threw a few stones;
No effect. Some more; nothing;
They could not halt
The foaming rabid beast.

The flight signal had reached his muscles,
Their fibers already contracted
To increase his running ability;
Heart and legs racing,
He ran and then looked back
To see the savage dog gaining on him...

He threw a larger stone;
It even hit the dog,
But there was no overall effect.

Graybeard reflected on
All his years on ToeQuest,
Wishing that he had said "boloney"
A few more times.

The crossroads all went nowhere;
The signposts pointed to oblivion.

The vicious dog was almost upon him,
So he stopped and waited
And planned for the fight,
Having but a second.

He wished that he
Had brought a weapon along;
There was not even a stick
Or a branch lying about.

He recalled his bevy of girlfriends,
But for the one he had given Austin,
For she was not much of a scientist.

Eternity called out Greg's name.
But this was a wrong number,
For he was now totally
Graybeard the Magnificent.

The foaming dog leapt for Graybeard;
Even one bite wold be fatal—
Graybeard's sturdy hiking boot
Caught the dog in the throat
As he kicked toward a vital area,
Stunning the dog
And sending him to the ground.

Just as the dog was about to recover,
Graybeard dropped a knee
Into his head and crushed it,
The poor creature's brains
Splattering all over the ground.

Greg's body and mind still swirled
With the rapid firings
Of the acute fear response,
But, he eventually calmed down.

There was a sour taste
In his mouth—
His salivary glands
Were turning back on,
A good sign that his life
Was returning to normal.

He walked back to camp
And drank the beer
That his glands had further requested.

Greg wondered how ninja Graybeard
Had accomplished the kill, then thought,
"Thank you, evolution and natural selection!
You made me what I am today."

Another dog arrived, a tame one.
Greg talked to it like a friend.

MUNDI EPILOG

'Twas the Pope's highest Cardinal [Sin] himself
Who'd ordered the assault on the Gray One,
And so the end was to be at hand,
But, on the other hand, Graybeard had algae,
And had flung it into the eyes of his pursuers,
Then patted the end of his horse.

The knights faltered and gave up the chase,
But the spiritual chasers appeared in their stead,
Stating unscientific theories,
Such as "God is".

And then?
The Spiritual Chasers of God Arrived.

*Meanwhile, science was (re)written
To say that some ancient wise men
Had long ago discovered
The weak and strong nuclear forces
While thinking about earth, air, fire and water,
Then correlating it to consciousness.*

*One of the ancient ones discovers a photon,
As well as the entire electromagnetic spectrum,
The strong force and the weak force,
Which corresponds to thought,
But they called it "Taurus".*

The spiritualists wanted it all,
And so they had to ordain themselves as "special",
Above and beyond all the rest,
For that way they could be deserving
Of even more reward in the afterlife,
All this born out of their pride
Of their very own Divine Creation...
That they made up.

Greg had befriended some of the spirituals,
But had to use logic on some of those remaining...

From the top of a large weathered rock,
Graybeard cried out, "Where is God?"

Each spiritual answered in turn:

"He is between our heartbeats and breaths."
"He is life."

"He is the universe."
"He is love."
"He is everywhere."
"All is of His illusion."

Graybeard answered,
"You have said nothing but life is life,
And that the universe is the universe, and so forth,
Just equating one real thing
With another name of an invisible thing
That is even quite undefined in the first place.

Who are you all that makest all of these words
Plied upon and on top of what is?"

"We are the spiritual chasers of God;
We label Him as anything and everything we choose."

Old Gray, looking a bit like God himself,
With his long gray and flowing beard, continued,
"Are you human mammals of such recent vintage
So extraordinarily important and special
In the whole entire scheme of things
That took so many tens of billions of years
To stumble along in such as haphazard way?"

"Yes."

"Do a trillion stars exist just to illumine your night?"

"Yes."

"Do forty million species thrive just for your delight?"

"Yes."

"And is all of space out there just for show, to glorify you?"

"Yes."

"Did the supernovae stardust showers
Of the atomic elements write the names
Of future humans across the sky way back when?"

"Yes."

"Does every atom exist and spin to service you?"

"Yes."
"Did Proto-men, and before them, and all, live,

Die and suffer only for your promise?"

"Yes."

"So, then, every dinosaur,
And more, was condemned
So you could gain a space to live, war and kill?"

"Yes."

"Does the sun shine with you in mind?"

"Yes."

"Was Heaven's Shrine built just to await your coming,
You being so special as to deserve a divine reward beyond?"

"Yes."

"Oh my, religious ones, how vain and proud you all are!
What hubris, conceit, self-love, and vanity
Have you to claim such full self-importance
To demand so much from the universe...

That you would even claim an angelic vapor that
Drives a living being, provides character,
Morality, and consciousness, on top of
A burdensome, fragile, and expensive
Organ such as a brain ne'er to be used?

It's a silliness born from exaggerated
Self-worth, an invisible hilarity—
Becoming a merciless indoctrination.

May you all soon recover your humility."

As such spoke the humble Graybeard to show
The truth of what we all are: mammal, organic;
Past narcissism and self-adulation,
To the bio-electro-chemical organism
Evolved upon a planet near a star,
In the long and winding mindless way of
Slow time, dust, and selection by death that
Sifts the best from the rest: evolution.

And such did Analog once observe that
The creature out thinks the creator,
With inferior tools, to imagine a
Much more peaceful and enjoyful world,
And that it is emotion that creates

Delusions of heavenly scenarios of
Creation, and an existence beyond death.

These are lessons in humility to all
Mammals grown so high and haughty...

So... enjoy it all as though you will never
Know it again; for who is to say that you shall?

Only one spiritual was left at the base of the rock,
But she had Graybeard surrounded...
With her words on evolution symbolized by the Bible.

"That's it," said Gray Newt, "I'm gone",
And so he fired up his jet pack
And launched himself into the sky
Toward the white patches of vapor and fluff.

Some of the aboriginals now thought him a God—
And so he became the legend—
That God was an old white-haired
[Gray bearded]
Guy sitting on a cloud.

MUNDI MEMORIES

Greg walked to the mountain and back,
Sitting safe on some lone rock for the lack
Of any other seat to pick but that of his own;
Wherewhence he slept, thinking he was all alone.

As there he lay asleep so peacefully in repose,
Some dogs wandered by and licked his nose.
And while he turned untossed, a kangaroo
Of boundless flight, hopped over him, too.

The Great Equalizer stalks all creatures made,
Lying ever just 'round the corner in the shade,
Taking both human and the beetle as one,
After their lives are spent from rolling some dung.

IN LIEU OF KNOWING THE TOE,

Some go off and imagine scenarios
That they think must be the TOE,
And further develop them
Into a 'concept of good',
Or at least a concept of
"This is it"
That works for them,
As "good" is a relative term.

When they see differences
In other cultures or TOE beliefs,
They label these ideologies as
"Not good",
Meaning not their own,
And thus they must be wrong
And 'evil',
And so many problems and wars begin
And then go on "forever".

Thus, their arbitrary and flawed 'goods'
Became the root of evil.

*If the false teeth
Of these false TOEs were exposed,
Would the world improve?*

Well, who knows,
For change takes a long, long time,
Especially in these matters.

THE UNCAUSED

There cannot be endless causes beneath causes;
Therefore, intuitive or not,
Something like the quantum doings
Are totally causeless,
Just as we observe that realm to be.

Cause is only of the subsequent realm.

It is only human to think of more
And more cause, ever deeper.

ON SCIENCE DISCUSSIONS

References to "outside" forces
May ever come as replies,
But, as always, if you use a word,
You must fully explain that word
In all its specifics,
And, in light of ToeQuest
And the Theory of Everything,
You must also do that
On scientifically satisfying grounds.

Mere pronouncements will, of course,
Only show up as such
And will appear to be undebatable
Just by the saying such of so;
As such, they are of no use.

"I think..." can indeed pronounce many things,
Many of those of an invisible realm,
But, again, this being ToeQuest,
The missing particulars and specifics
Don't help the claims proceed along...

Thinking of 'God' exhaling/inhaling,
Although still a wonderful thought,
Raises the question of the evolution of "God"
And the further speculation
Of who and why He must be.

Similarly not conducive to centering debate,
Are such diversions done before.
As painting Darwin as a God
To be followed blindly,
As Gods are,
Or of evolution being akin
To religious dogma,
Which it is not at all, since it has facts,
Or, having evolution be real,
But then again not real,
For one's other purposes,
Would be off the mark of the discussion
Of the actual topic of at least
The posts on evolution,
Which, is course,
What some may wish for.

I understand if some are adverse
To debating the visible realm of existence;

But, for ToeQuest, this is the way
For these discussions to go.

Of course the ideas of evolution
Are often used cosmologically, too,
As for what succeeded in this universe
Or among many of the multiverse,
As well as in many others disciplines.

EVOLUTION INTRODUCTION

Charles Darwin,
Working long before
The DNA/genetics revelations
Of the 1950's,
Was a Victorian era scientist
Who constructed his revolutionary theory
Of evolution through natural selection
Over a lifetime of meticulous observation and thought.

It is perhaps the must powerful idea in science
And still drives the contemporary research agenda.

Life was staggeringly dull
A billion and a half years ago,
And, of course,
In the 3 billion years before that,
The once steaming ocean having become a cold,
Thin, dreary broth of look-alike organisms.

Eukaryotic cells
With internal structures had appeared,
But not yet were there
Any multicellular creatures.

Life lazed through those doldrums
For a million millennia.

Imagine the lengths of these times.

Then some combination
Of environmental circumstance
And genetic novelty triggered
A wild diversification in
The variety and complexity of animal life
Over tens of millions of years,
Climaxing in what is now called
The Cambrian explosion.

By 530 million years ago
The seas held a multitude
Of bizarre creatures—
As now seen fossilized
In the Burgess Shale,
Perhaps near where
Graham of ToeQuest lives.

As is often the case,
Many of those weird Cambrian monsters
Were evolutionary dead ends,
But a certain few were
The progenitors of every animal alive today.

Darwin's real breakthrough
Was that evolution became inevitable,
Since, in organisms whose environment had changed,
Those who had reproductive success
Depended on inherited traits.

Then, too, there was the simple mechanism
Of natural selection,
Although there could be more methods.

Since then, Darwin's ideas have connected up
With genetics, molecular biology,
Ecology and embryology.

Today, Darwin's legacy is a larger,
More richer, more diverse set of theories
Than he could have imagined.

While the competition
For ecological resources,
Also called natural selection
Or "survival of the fittest",
Demonstrably drives much
Of evolution and speciation,
Biologists are now onto
The elaboration of Darwin's ideas
About sexual selection,
Plus ongoing debates about roles
For selection at the single level of genes,
The individual organism,
Whole species, and/or all of the above.

A CONSPIRACY?

Darwin was a puppet on the end of puppet strings,
Yanked and pulled by those with the power
And know-how to manipulate the social environment
And the forces of the masses....

This one was too good to pass up.

So, then, some nefarious planners
Set out to demean humanity
By knowing of Darwin's results ahead of time,
And so they sent him on a five year trip
Around the world on the Beagle ship,
During which he actually found adaptation
Of the finches, and all, as input to his theory.

Then, luckily, the rock strata were dated
And all the fossils were in the right strata
For their place on the evolutionary ladder,
We also noting that some species went extinct,
Not appearing in any more strata,
And that some new species appeared above,
Along with some transitional forms in between,
All this proving Darwin's theory,
Not to mention science seeing evolution in action
Via 40,000 generations of bacteria in the lab,
Plus discovering the genetics behind all this,
This DNA correlating with the fossils.

Then the nefarious planners widely publish,
Then, and now, and at times
So that the actual truth
Can be learned.

What a shady scheme!

MORTALITY

We are chained to Time—
To the speed of light,
A fair price
In return for 'existence';
But, the timeless-formless
Potential is free
To conceive all there could be,
'Possibly'.

THE 50-STEP PROGRAM

1.
My god is obvious.

No — it's certainly not obvious,
But it just seems so because it is a beautiful wish.

There are even many tragedies in life
Which make it appear that, 'obviously',
A benign being can not possibly exist.

2.
Almost everybody on Earth is religious.

Not anymore.

In history, humans do seem to
Have religion throughout the earth,
But, with the growth of scientific knowledge,
This has become much less so.

3.
Faith is a good thing.

Faith is necessary to some people,
But is neither a good nor a bad thing....
Just thought necessary; otherwise,
One would be so full of fear
That one would not be able to function;
But, then, too, arrives the fear of Hell.

4.
Archaeological discoveries prove that my god exists.

Archaeological discoveries only prove
That religion has existed....

5.
Only my god can make me feel significant.

Because one is conscious
One knows one is significant,
But not necessarily important.

6.
Atheism is just another religion.

Atheism is another way of being....
Of looking at life.

7.
Evolution is bad.

Evolution is a way of looking at
What has happened—it is neither good nor bad.

8.
Our world is too beautiful to be an accident.

Our World is very beautiful.....
But very cruel, and in many ways corrupt.
The world and the universe at large
Are not congenial to life.

9.
My god created the universe.

The forever stuff, the basis,
Never had any definition,
For there was no time at which
It could have been defined.

10.
Believing in my god makes me happy.

Believing in no beginning and no end
Makes me feel 'satisfied',
But, sometimes I am happy
And sometimes unhappy;
This is just the human condition.

11.
Better safe than sorry.

This is an obnoxious philosophy—
Honoring 'God' with lips
But hearts far from seeking
The truth/enlightenment.

12.
A sacred book proves my god is real.

The sacred books cause most of the problems!
A book is always only ever about men's thoughts....

13.
Divine justice proves my god is real.

There doesn't seem to be much justice,
So, that can't prove anything.

14.
My god answers prayers.

Amputees are never healed.

15.
I would rather worship my god than the devil.

This assumes that there is
A Good Being and a Bad Being
And that either of them
Would be small-minded enough
To want to be worshipped;
A feeble assumption.

16.
My god heals sick people.

If he did, he would heal them all, not just some.

Prayer can help a person to cope
With pain and suffering....
One's own or another's.

Prayer/meditation/contemplation
Can relax a person
So that the body can be helped.

17.
Anything is better than being an atheist.

It is OK to be an atheist—
So long as you don't wish to exterminate

All the ones who are agnostic
Or just merely just hopeful
That we are not merely just flesh and blood.

Anything is better than a person
Who wants to kill another
Because he doesn't perceive life
In the same way.

18.
My god made the human body.

The human body evolved.

19.
My god sacrificed his only son for me.

No.... because the Christian faith,
Which gives this message
Also says that the son is still alive....
So, he's not been sacrificed, as in exterminated.
Nor did the Jewish Jesus
Even intend to start a new religion
Based on this or anything

20.
Atheists are jerks who think they know everything.

They know much more than what is simply pronounced.

31.
Intelligent design proves my god is real.

The 'design' is not intelligent,
But, if you insist,
Then God had to be designed, too.

32.
Millions of people can't be wrong about my religion.

Millions of people can't even possibly agree
About what constitutes religion.

Two people can't even agree entirely....

33.
Miracles prove my god is real.

Depends how you define a miracle.
An empty chrysalis which becomes a butterfly....
Is delightful. This but proves that nature is real.

34.
Religion is beautiful.

Religion can be very ugly indeed.
Truth is beautiful.

35.
Some very smart people believe in my god.

And some very smart people don't.

36.
Ancient prophecies prove my god exists.

I haven't seen any ancient prophecies
That foretold anything.

Even history is, sometimes,
Warped to fit what the narrator
Wants you to hear.

37.
No one has ever disproved the existence of my god.

Austin has.

The Theity who is supposed
To be everywhere cannot be found;
Only the natural is found.

Even a Deity could not exist
As the first and eternally causeless fundamental.

This is because a system of mind
That can do planning, design, creating, and so forth
Would have to reply on parts
That were even more fundamental;

Plus, that the complexity of being and other composites

Lies way at the other end of the spectrum
From the fundamentals;

And that nothing comes from Nothing
And so there couldn't have been
Any creation or Creator of the ground-state,
It having always been there
As the natural state of affairs.

38.
People have gone to heaven and returned.

Where are they?

There is no such place as heaven...
There is just a state of mind.

39.
Religion brings people together.

Religion separates people.
It becomes 'they' versus 'us'
Because different people
Have different perceptions
About what is important.

It is OK to be separate.
It is when we all pretend
To be nodding in agreement
That the trouble begins.

40.
My god inspires people.

What is in people, not God,
Can inspire them....
As artists, writers, musicians....
Whatever.

41.
Science can't explain everything.

Not yet, but it does more and more.
We don't know what 'black holes' are, do we?
We don't know what anti-matter is do we?
But we know they are there.

We haven't explained them yet.

There is plenty of time though.

42.
Society would fall apart without religion.

Society would fall apart without laws.
Religion, too, has been used
To make people abide by the laws.

43.
My religion is so old, it must be true.

On the grand scale...
No religion is very old.
Humans haven't really been here
On the earth for very long.

We are still evolving...
If we don't exterminate ourselves first.

44.
Someone I trust told me that my god is real.

What people have told me
Has made it all seem 'unreal' at times,
Not to say downright silly.

45.
Atheism is a negative and empty philosophy.

Atheism is just a state of being.
Perhaps one needs to reject 'taught' religion
As silly before an inner god enlightens you,
If and when you want to be enlightened.

46.
Believing in a god doesn't hurt anyone.

Believing and trying to impose
Your belief on others can hurt millions.
Otherwise, fine.

47.
The earth is perfectly tuned to support life.

Earth is perfectly tuned to support
Some forms of life unless
We botch it all up completely.
Other places definitely are not.

We have been tuned, via evolution,
For our continuing life on earth.

48.
Believing is natural so my god must be real.

The search for a satisfying spiritual life
Can be 'natural' from natural selection.

49.
The end is near.

Says who, 2012?

50.
I am afraid of not believing.

Yes... I think, therefore I am...

If I stop thinking I might stop being...
And in spite of everything...
I quite like being.

**WHAT ABOUT THOSE PEOPLE
WHO ALWAYS SAY THAT
NOTHING IS REAL,
SUCH AS ATOMS
AND EVERYTHING MADE OF THEM?**

Well, then there ain't no H_2O, nor enzymes,
But the funny thing is
That the atoms of the 'dream' H_2O
And the 'dream' enzymes and so forth
Work 100% the same as if they were real.
What a coincidence!

IN THE BEGINNING...

At an arbitrary point in time
God obtained some material somewhere,
Stuck with its limitations,
And thus created
The Heavens and the Earth.

Eve then cost Adam a rib,
And later on an arm and a leg.

Cain killed Abel to have sex
With Eve or his sister(s)
Since it was so hard to multiply
While being fruitful.

Cain killed Abel, and, so
We are all Cain's children.

So, our ancestors descended
Not from the trees,
But from God's
Evil human nature Design.

Noah married Joan of Arc and then took her
And all their pets on a world cruise,
Noting the rest of the human race
As dead and drowned.

God played a big joke on Abraham,
Whose kind had often made burnt offerings
When popping the corn or
Overcooking the Lamb of God.

Moses then tied his ass to a tree
And wandered away to cleanse the tribes,
Then lost his way for 40 years,
Not even stopping to ask directions.

The ancient Egyptians fleeced the electrolytes
While the Futurians wrote many letters.

God made spiritual love to a teen-age virgin,
And Jesus was born Jewish,
Then converted to Christianity,
Died, and was born-again
On Easter (Let us Raise the Lord!);

But not before Jesus
Had made water into wine,
Perhaps encouraging alcoholism.

Mass was served by the altered boys
And even odd girls, all then preyed upon.

Lent soon became fast-food only time
And so Fat Tuesday was invented
To tide one over.

Every Friday the priests had nun.

God be in my heart,
My mind, and my end.

Thank God!

Sleep be with you.
A-choo.

HOME SWEET HOME

Some people drive treacherous roads
Up to a mountain top
When it is snowing so they can ski,
Paying a lot of money
For the privilege of breaking their bones.

They must have poor mental health,
For they even put slippery wax on their skis
So they can go even faster
Down a mountainside
That is already slippery
And full of snow and ice and trees.

These humans have not evolved much
Since the time of the Woolly Mammoth.

TOUGH LOVE

This insulting I would not do,
For many eyes may be upon you.
And, yes, we do care
That you are there;
If such love seems rough
It is but the kind of tough.

THE ITCHY CONVENTION

Many of the claimers of absolute knowledge
Of the workings of the invisible unknown
Had been gathered into a conference room,
For here they would be face to face
With all the variant and differing
'Proved' beliefs of ultimate truth
That were so indubitable
To each of their sponsors

Here there would be none of
The 'neglect' of contradictions,
For while the individual beliefs
Were very personal,
And thus unassailable—
Since they had each
Merely thought of them,
There would be unavoidable debate.

We knew, too, that they would all get angry,
And so, to accelerate the process,
Itching powder had been sprayed
Into the air beforehand.

"There are no gods, just ways of life
Indicated by the consciousness of the universe."

"Nope, there is one God."

"Ha, hardly. There are many Gods,
Krishna just being of of the better known ones."

Wham! "Go eat a sacred cow!"

Baam! "There is no Heaven or Hell."

Whammo! "Have a nice warm trip there."

"My all loving God wouldn't torture anyone in Hell;
That's more like a Devil would do."

Punch! "Don't call my God a Devil."

"No change to what 'is' is needed."

"What! One must greatly reduce
Or banish the ego altogether! It is Satan."

"Well, can't use an ego to banish an ego."

"The ego will surely be gone after death."

"But we never die; we are an ongoing dream."

"OK, never mind; I love you, Mel."

"Jesus was Divine, fat Jewish head."

"No way, Islamic monster;
You stole that concept."

"Death to you, infidel.
Mohammed is the main man,
For an Angel told him everything."

"God talked to me and said
That the Book of Mormon
Is the true faith."

DAILY PRAYER

Our Father, Art, in Heaven,
Harold be thy name on Earth,
Hollow be thy fame.

Our will is done, so, someday,
You can blow us to kingdom come!

But, first,
Give us this day
Our daily news, e-mail,
Texts, UPS, and Fed-Ex,
And our PMs and posts,
And, please,
Forgive us our trespasses,
For the only thing
We cannot resist
Is temptation;

However, do not soon forgive
Those tele-marketers and door-knockers
Who solicit and trespass against all of us,
But do deliver us all
Unto the pleasant 'evils':
Whether a-men, a woman,
Or whatever the preference...

RELIGIOUS ANGER

In a galaxy, among trillions,
In the middle of nowhere,
On a planet, among billions
Apparently not needed,
Called Earth,
Out on an arm
Of the Milky Way Spiral,
Some organisms formed
Over five billion years,
Leading to over 30 million species,
Among them some higher mammals,
Such as dolphins, chimps, and humans.

These mere specks of life then decided
That they were special
And part of a grandiose scheme.

Yes, the human mammals came to believe
That they were much more than
Just a natural outgrowth
Of the organic world—
That they were placed here
For a special purpose.

They claimed that God's imprint
Could be see all over nature.

Someone asked "Where?"
Then they got all mad.

ALL SUMS TO THE NONE

The timeless-formless contains every path,
Though as useless as a library of ALL books;
For, its sum of information is zero,
But, one of these possible avenues became ours.

NO-WHERE AND NOW-HERE

We shine as un/real and shimmering rainbows,
The stable/virtual still the same as the real;
This differentiation was of potential—
Being there's no way to have Nothing.

SELF-CONTAINED

I'm using the term 'Universe'
Since it is all, complete within itself,
Anything outside being, well, outside.

The Universe just keeps on Universing
Of its own accord, of its own energy that it has.

If someone wants Someone Else
That was 'just' made of Itself,
Then I, too, take that,
Even all the more,
For the much lessor thing
Of the Universe made of itself.

Higher management [God]
Has been laid off,
Apparently, for it was never there.

Is the Universe the same as God?

Well, no, since it matches only
About 1% of the 'God' that people
Have come to believe in
(The made from itself part,
Plus the awe for such a great thing).

So, we really need a better name than 'God'
For 'what is' until people
Come to see 'what is' as being itself.

A bridge for now could be
The Ground of Determination (G-O-D),
But, then, after that,
We can just say 'what is'.

We might even all stand up
When the word 'what is' is said.

WHERE NO PRIEST HAS GONE BEFORE

Her legs beckoned the great beyond('s) delight...
But, lax in psalm reading, the priest had none,
Her paradise lost by the dashboard light;
But, next time, of heavenly body, he'll have nun.

MEDITATION—
THE STATE OF WHICH
IS NOT WHAT YOU THINK

Calming one's mind
By removing all thoughts,
Via meditation, results in
A pervading neurological sensory feeling;
Yes, it is still a felt sensation.

Some might arbitrarily equate
This to "God" or the "All";
Yet, it is only a feeling
Induced by meditation.

Simply pronouncing it
To be something else
Is to fool one's self,
For, a sensory feeling
Is yet a sensory feeling.

Lab testing of Buddhist monks indicates
That certain neurological areas
Of the brain indeed go quiet
And stop reporting
About the self and its boundaries.

I have had the same feelings
Through healthful meditation
That others have had,
But, I do not claim the 'sure things'
That I am floating in outer space
Or going to other realm,
Even though it feels like it,
Nor do I claim that
I have merged with the chair,
The room, the world, or the universe,
Even though it feels like it,
Nor do I claim all is nothing,
Since that's how emptiness feels.

Nor any claims of "God", nor "All",
Nor merging with the Cosmos.

It is just a neurological feeling that's
Especially induced by the quieting
Of brain areas for the self
And the body's boundaries.

The feeling comes from one's own brain;

It is not an external signal sent
So we can touch God, know God,
Or peek into God's realm,
For this is all just wishful thinking.

Here is more:

Humans have become suckers
For many weird beliefs.
They have even been encouraged
By brain systems that evolved for other things.

For example, as in meditation,
A bundle of neurons in the superior parietal lobe,
A region toward the top and rear of the brain,
Distinguishes where your body ends
And the rest of the world begins,
This being quite a useful feature.
Without it, we would bump into everything.

However, brain areas send "turn off" signals
To this area when we are falling asleep,
Having sex, praying, chanting,
Or meditating deeply.

One then feels part of something
Larger than one's self; however, one is not.

When the brain is unable to find
The dividing line between self and world,
The brain adapts by experiencing
A sense of holism and connectedness
That then promotes other beliefs
Which bring a further sense of connection.

The quietus of these brain areas
Lead some to figure that
They have become one with the cosmos
Or with God during meditation;

But, alas, it is only
That induced floating feeling
Of oneness, still a felt sensation
In the 'state of being'
That ever resides atop the mind's
Unreachable chemical-electrical state beneath.

Who then, upon recalling,
Or during a state of meditation,
Or just by living,

Can use experience
Of doubtful analysis
By the mere state of being
To say anything further
About the true nature,
Why, source, how, and wherefore
Of the conscious awareness
Of felt sensation
Without any consideration of the
Electrochemical states beneath?

There are those who feel
They really have to say,
Based on introspection alone,
And those who do not,
For science informs us
Of the states beneath.

TODAY

The day was yet glorious,
The leaves of the trees
Undulating from the wind,
And then they themselves
Making another smaller wind
From their movements.

I breathed the oxygen
That was a waste product
That bacteria had given off
And walked in my chemical nature
Where dinosaurs had once tread.

I felt the life, the awe,
The triumph and gladness,
Even the glorious defeats and tragedies;
But, life ever goes on,
And it was filled with
The love and adventure
Of living this expression
Of the Universe.

INFINITE/FINITE/NOTHING

The universe is finite,
And there is Nothing beyond it;
The Nothing is either finite or infinite.

We have to remember
That Nothing has no existence,
Meaning that it is not there;
Thus it has no properties,
As do things that are there.

The finite universe ends where its influence ends.
Nothing does not just sit around as empty space;
It can't, for its meaning is 'not to be'.

Here's Nothing! See it?
No, for it is not there.

Also, the "infinite" is what can never be reached,
By definition, so, there is no reaching it.

Potential infinities are like
The endless counting of numbers on and on.

Actual infinities are like imagining the subdividing
Of a finite line length into endless points
That are yet still contained within the finite line.

So, neither Nothing nor infinity can be actual.

SHAMAN YOU

You shame the sham of even the Shaman,
Time and time again with rudeness, man,
Only very few of these times getting banned;
This is not for the TOE to grow as science land.

Not liking scientific posts on this place meant,
You bash, burn, and swear out of ego's tent.
This is all because you can't answer the content,
It being rational and logically not of your bent.

Religious preaching, as you even well know,
Though you love it so, is prohibited on the TOE.

THINKING

Perhaps we could loosen
The definition of a thought
So that it is just not
Of a system of mind,
For that would really need
Its own prior creation from parts.

As long as a "thought"
Finds a solution,
It could just as well be
The brute forceness of something like
All quantum-type paths being
And evolving in superposition.

We must preclude anything
Real and defined,
Such as Intelligence,
From having been around forever
Without ever having been defined
In the first place that never even was.

Thus, there is always a "before",
But only to further complexity
Or even the simplest structure.

This last "before" is the state
Of Potential or Possibility
And thus it could surely
Have been around "forever",
Although it is more of a timeless,
Formless, thoughtless,
And lawless nature, to us forms.

After this secondary kind of 'creation',
Then time, form, thoughts,
And laws came into play,
At least for that which became.

How come Nothing didn't dominate,
Leaving nothing "here"?

Well, who knows if it even could 'be',
But at least this shows that something had to be,
Since there is, and so we are relieved
Of explaining that question.

The state of all universes went everywhere,
They evolving on in this superpositional state

Until some mammal or other species' consciousness
Real-ized and actual-ized the universe,
It then continuing on as real.

This is all deduced because
There had to be "something" around forever,
Although it could not have been a real thing,
As the defined requires definition,
Not to mention that an eternity
Can't have already completed.

The every-path approach gets rid of having to have
A predictive system of mind of a Being
To explain doing the planning
Of 10**kazillion interactions and so forth,
As well as it being a more surefire method
That can't lose—
Since no stone or path is left unturned.

Again, it would have been a problem
For a real thing to have been around forever,
In that a forever could not have already passed,
As, by definition, infinities can never complete,
At least those that are tied to time.

OK, that's the TOE,
But, our existence, of course,
Is what we have to deal with,
So who cares about our essence,
The details of which can
Never really be fully known
Since no one was around
At the origin of the universe.

Thus, we are free to be.

GALAXIES HATCHING

Hawking radiation may cause
The evaporation of a black hole,
But what of this radiation?

Could it be that black holes
Within a galaxy
Actually spawned the galaxy
In which they reside,
The egg arriving before the chicken?

THE WISH

"Who wants to be a part
Of something larger than
A whole universe developed
Over 13.75284129 billions years?"

"I do," said a passerby.

"How come?"

*"Then I will be looked after
As the special
And super-important person
That I hope I am."*

"Sounds good if it can be so,
Although it be not so humble."

*"The immaterial soul makes me important
Since it can receive outside direction
And can even continue on after death;
I deserve an infinite reward for all eternity."*

"'Soul' stands for
'Spirit of Unconditional Love'."

"You invented that?"

"Yes, and I led
An anonymous group called that."

"What happened?"

"They ended up giving love all over the place,
Right and left, up and down,
And fore and aft,
And then felt really great
About it and themselves."

*"So maybe there is a Higher Power
Like that who made us."*

"Does S/he wish to be a part
Of something larger, too?"

*"Oops. I don't know,
But let's say that there is nothing larger."*

"OK, then; so now you are a part
Of something larger
And will live on forever,
So you can just relax and enjoy life."

"That's good news,
But I wish I had proof."

"Well, so-and-so wrote or said
That there is a soul."

"I have to see evidence of it.
People say lots of things."

"OK, let me know."

"I found no evidence; no God."

"Amen."

VISUALS

The visual system uses a lot of
Real estate in the brain,
Having neurons for locating edges,
Sensing shadows of them even,
And more than we can say here.

That we have two eyes,
Each reporting to
The opposite half of the brain
Allows for knowing distance.

I suppose it's really all how and where
The photons fall on the retina
That gets ever more interpreted
By higher levels.

An unseen but somehow felt
Really fast black and white image
Gets sent, too,
So we can quickly duck away
From a threat.

LAST, NOT FIRST

Consciousness is not a thing
Floating around by itself like a soul.

Consciousness is not
Even a thing in itself,
Or some container
That results are poured into.

It is the tail end of a process
Of what the brain does,
A resultant global brain state
Of the brain's subconscious analysis.

Our environment, inside and out,
Is symbolically represented in the brain,
Our memories and cross-associations
Recognizing and remembering the meaning
Of what we "see",
Think, feel, and witness
In the unified experience
Of living life as a being.

We all know how wonderful
And quick the brain is,
It nearly instantly processing visuals,
Sounds, touches, tastes,
Odours, and feelings
Into higher and deeper sub systems,

And how it searches memory
So quickly for what is known,
Such as what the letters and words mean,
Then forming an abundance
Of further thoughts and actions
Based thereupon, and so forth and so on,
One hundred billions brain cells
Winking and blinking, and connecting,
Making their results known consciously,
At the last, this continuing on
In a further train of thoughts
Ever becoming and arising
In a kind of competition for attention.

Some might propose
That this does not happen,
Even that the brain does little or nothing,
Or that it suddenly hands off its information
In mid-stride to "somewhere else",

Called the "soul", "mind" or "consciousness"
That somehow continues the processing
By using some outside, invisible guidance.

Ask where is the soul
And no one knows,
For it is invisible.

What does it do
That the brain doesn't?
No one knows.

How does it attach to a person?
No one knows.

When does it attach?
No one knows.

Does it have eternal life?
Many certainly wish for it.

How does it operate?
No one knows.

What sent it?
They say that they know.

How?
No one knows.

How can it interface back and forth
With the material brain
Without being of the material
And speaking that language?
No one knows.

Why is it there?
The reasons are made up.

The extraordinary claim of the soul
Has no evidence at all,
While the extraordinary claims
Of the brain's doings and its
Being the same as the mind
Have extraordinary evidence.

Yet, some mammals continue
To employ the world "soul"
Because they wish it to be "something"
That makes them so absolutely special

And of such extreme importance
That they must deserve
An infinite reward
Of eternal life forever
In the most ultimate place.

What hubris!
Not just any hubris,
But the maximum amount of hubris
That is possible for the never so humble:
All.

BEING

'I'
(Awareness/Consciousness,
Still and always of the brain)
Observes
(as a witness)

The state of mind
(experience of thought
Or feeling or sense, etc.)

That has surfaced
(the brain's result of analysis)

From subconscious brain scenarios
(associations, learnings, memory,
That is, the self)

Due to this situation of inputs above.

(That's life.)

TALENT

Isaac Asimov said to all, "Don't thank me,
For I had creativity naturally.
I loved science and so I drank it in;
All that came flowing out was within."

DREAM LIFE

Dreams demonstrate the true excitement
Of every night life during the time of sleep,
An even more wondrous time
If you're lucidly aware in the dreams.

It's like a movie filmed in Cinemascope and 3-D,
A virtual reality in which you can script and star.

I will be making video clips of some dreams
By having an electrodes implanted in my head
And will then have Steven Speilberg or George Lucas
Splice them together into a movie."

Last night's dream was that
I was in downtown Poughkeepsie,
Not the greatest place at night,
And had to take a taxi,
Which proceeded to drive over curbs,
(The bump was not felt as much as it should have been),
Up big hills, and over really impassable terrain,
All with great ease.

I was only half lucid, I guess,
And so I thought that it must be
Some virtual reality amusement park ride.

There were some really glorious curving icicles
Atop the stone hills, tusk shaped,
Something not really possible.

So, is that experience a part of me now?

PLANT "LIFE"

It is true that plant's awareness
Is very rudimentary,
But their cells do sense
(In a chemical way)
Light, heat, foreign cells,
PH condition in liquids,
And many other states of matter
That can be good or bad for their survival.

MENTAL ILLS

Seriously anxious and/or depressed people
Can get aggressive and lose friends, lives,
Marriages and jobs and so forth.

Obsessives are bothered night and day
By the persistence of
Very intrusive and repetitive thoughts,
Often falling for them hook, line, and sinker.

They might wash their hands 50 times a day
And/or come home from work
To make sure the coffee maker is turned off.
They might even bring the coffee maker to work
So they don't have to go home and check it.

ADD causes one to miss the main event,
Being distracted by things that don't matter so much,
Giving them way too much attention,
Even, like a cloud shadow going by;
However, they can't much help it,
For a wrong signal goes off,
More like a siren or an alarm bell,
That says "ATTENTION:
This is VERY IMPORTANT".

Meanwhile, the main event of living passes by,
Piles and clutters of stuff
Usually building up around the home;
They can get angry if others don't see the importance
That they claim about certain things.

Next time: the trade off between drug side effects
And the desire to feel normal and be able to live life.

Serotonin is a "biggie" in psychiatry,
As low levels of it are the root
Of obsession, depression, and anxiety.

Serotonin is a kind of traffic director
Of the brain's signals.
And of course there is more
To the resultant cascade.

In the central nervous system,
Serotonin plays an important role
As a neurotransmitter in the modulation
Of anger, aggression, body temperature, mood,
Sleep, human sexuality, appetite, and metabolism.

Serotonin has broad activities in the brain,
And genetic variation in serotonin receptors
And the serotonin transporter,
Which facilitates reuptake of serotonin into pre-synapses,
Have been implicated in neurological diseases,
As well as in the resultant lifestyles.

Drugs targeting serotonin-induced pathways
Are being used in the treatment
Of many psychiatric disorders.

However, although drugs target
A specific serotonin receptor,
Inhibiting its reuptake
(And thus providing more of it),
The other types of serotonin receptors
For things like sex and sleep are so similar
That they can often be affected
By the drugs, as well (these are side effects).

It's up to the patient to decide
If the "cure" outweighs the side effects.

To help with insomnia,
One could take the pill in the morning
Instead of at night.

The sex dampening feelings may go away,
With enough persistence,
Or one may choose to live with it
As a much lessor of two "evils".

Cognitive therapy is needed, as well,
To have a multi-approach.

Also, since depressed, anxious people
Are already prone to suicide,
They may still do it,
But, some don't have the "oomph"
To carry through on it at first.

As some drugs take 3-5 weeks to work,
The drugs may provide the "oomph"
Before they provide the feeling of normalcy,
So, the patients beginning drug treatment
Must be carefully watched.

Yes, life is often a compromise.

INFINITE REGRESS

A particle with no ultimate source
Would take forever to act,
As it would have to wait for the influence
Of its earliest part to travel on up,
Of which there isn't any in the first place
If there is no first place going forever back.

In general, if say, energy were infinite,
Which, by definition, can't happen,
For infinities never complete,
It would be very crowded around here.

For the same reason of
No things having been around forever,
Since a "forever" could not have
Already completely happened,
I would say that real and defined things
Could not have been here forever.

Nor could things have arisen from nothing,
So, that leaves us with "something" else
Called Possibility or Potential,
Which is neither nothing nor a thing.

So, then, we are happy now
That there must be something,
As we are in it and of it.

"Infinite nothing" would then be
An endlessly incomplete not-anything-at-all,
A big zero going nowhere.

GENETICALLY MODIFIED FOOD

At breakfast today,
My ants spelled out SPECIAL K Blueberry cereal.

GM, a conspiracy run by a General named Mills,
Is a terrorist organization out to wither us from within;
These cereal killers will stop at nothing
To achieve their aims of killing mice and men.

Get that train of thought out of your caboose
And see the light at the end of the tunnel.

MATH AND REALITY

It could be that math,
Although it is a symbolic language,
Underlies all reality in a way,
Or even all the way.

It is seen that some of the equations
Describing the physics of physical events
Have been discovered before the events themselves
Are found and confirmed.

Nature must have a regularity
That lends itself to description by math.

I not sure if math handles infinities directly
Or uses some short cuts like calculus
That merely approach them.

IN THE ZONE

This universe, and the earth,
Seems to work within some tolerance.

If some stuff was different,
Perhaps the whole earth
Would have been a huge flop.

So, ours is a particularly rare
(At least for us and considering
A huge amount of failures elsewhere)
And workable universe, at least for our earth,
One that we all know and usually love.

If there are some things
That we don't like about earth,
Then it's still probable that,
They, too, were necessary
For the whole to operate.

No one likes worms and bacteria a whole lot,
But they aerate the soil so that plants can grow,
Not to mention that bacteria
Filled up the atmosphere with oxygen,
Which was nothing but
A waste product to them—a poison.

INFINITY/POTENTIAL/"SOMETHING"
HAD TO BE,
PRECISELY BECAUSE NOTHING CAN'T BE.

We have to remember that
Nothing means 'no-thing',
'No where', 'no how', etc.,
And so we can't counter its own definition
By having it do or make something out of itself,
For it has no self, no place, and no being.

As its definition also means 'not to be',
It cannot then ever be nor have any power.

It's like the square circle idea as a contradiction.

Nothing can ever become of Nothing
Because there's nothing to work with.

Even if something appeared
Where there was not something before,
It was put there from where
It was—somewhere else.

It's not manufactured on the spot by itself
With no outside help,
But with the power and energy
Of "something" operating to put it there.

That said,
And being careful not to refer
To Nothing as something,
Could it have been the case
That there was not anything at all
Anywhere in any way,
Even the "anything" of Infinity/Potential?

Nope. *Why?*

Because there is something around,
As we can see; so, for whatever reason,
A complete state of no "things"
Could not dominate the entire scene,
For what could have maintained "it"?.

As such, then we have an answer
To people who might ask
"Why is there something?"

There had to be "something" and we are much relieved
Of having to account for it's being;
So, it's not like "something" was an optional condition.

PATHS

I should also add to the TOE
That the preciseness of some
Of the constants in our universe
Shows only that many arrangements
Of universes from Infinity/Potential
Could "stumble" upon this rare
And quite workable arrangement,
At least upon the small Earth.

Not really "stumble", as in an accident,
For our universe's formula
Was guaranteed to be found,
Since we are here,
And all possible paths were ongoing

Now, as soon as I say this, I have to qualify it.

Infinity is not exhaustive, strange as it seems to say.

Pi, for example, 3.14...
Goes on forever, but if we, say,
Let some digits somehow stand
For letters of the alphabet,
Not all books would appear
In the infinite expansion.

What did Infinity/Potential
Make actual things out of?

Itself, for it was not a "nothing",
But a "something"/capability,
The only kind of "something"
That wouldn't need anything
Before it or beneath it,
For that would only be
The same kind of stuff
Of Possibility and Potential.

Any other TOE,
Such as Atlas standing on a turtle,
And that turtle standing on another turtle
Would need turtles all the way down...

JACK HOHNER'S 'DYNAMIC MATTER THEORY'

Jack, the 'Dynamic Matter' pdf on your site
Was one of the greatest
And convincing expositions
That I've ever read,
Resolving several "unresolvable" areas of physics;
You must have been delighted
When all the math worked out.

There is your simple but ingenious insight
Of having real particles accumulating
From Hawking Radiation even amounting to stars.

And the truth of why the curvature of space is not so,
Plus what is the replacement for the effects
Of what we thought was from dark matter,
How time is of a more consistent movement
Than the vague 'motion' commonly thought of,
Wave functions going all the way up,
And even why dinosaurs didn't collapse
From their own mass;
You solved a vast amount of dilemmas, Jack.

I'll leave the details for inquisitive readers to find.

Jack, I forgot to say that I suppose the TOE
(After the origin of the universe)
Is probably the proved unification
Or separation of the forces,
Which seems to foremostly need gravity
To be greatly clarified, which you do account for;
So, I feel that you do have a TOE in the regular sense.

As for the TOE of why and how
The universe appeared and from what,
Which I like to ponder, we need a special name,
Perhaps the SUPER TOE?

Do you think it is that
The influence of the universe
Is just potentially infinite,
Since the self-regenerating electromagnetic waves
Are still on their way toward 'forever'?

So, then, wherever their influence presently ends,
That is the edge of the universe,
With truly Nothing (that is, as in not there) 'beyond'?

Now, for a SUPER TOE type question:
What defined the [limited] amount
Of the finite energy of the Universe?

And, if the answer is that it was always around,
How did its amount, nature, and so forth
Get defined without ever having
Been defined in the first place
(There being no first place, of course)?

I can give you a minute or two
To come up with the answer...
Ha-ha on us all!

Could it be something
Like in the quantum realm
But even more than that,
For the quantum has energy,
In which there is
Neither nothing nor something,
Which I might call
The Possibility or Potential state
For a lack of a name
For this new [proposed] state
That emits particles and antiparticles in balance?

Possibility would need not anything before it,
For that would only amount to more Possibility.

SPEEDING AROUND

If I traveled faster than light,
Then it would be dark there,
Unless I brought a flashlight,
But that would probably
Look dark to me, too.

So, how about if I just be lazy
And sit around on the sofa?

This hectic world really has to slow down.

SCHIZOPHRENIA

Some would love to think that schizophrenia
Is somehow an integration of consciousness,
But it is rather a disintegration.

The severe cases cannot even function.
Schizophrenia is marked
By severely impaired reasoning
And emotional instability,
And can cause violent behavior.

It is a serious mental disorder
That affects millions of people worldwide.

By some estimates, 1-2 percent
Of the world's population may be schizophrenic;
People diagnosed with schizophrenia
Make up about half of all patients
In psychiatric hospitals and may occupy
As many as one quarter of the world's hospital beds.

Research shows that the condition
Tends to run in families;
A person with schizophrenic relatives
Is ten times as likely to develop schizophrenia
As someone who has no history
Of the disease in the family.

One hears voices that
Are not considered to be one's own,
Such as would be so in a dream
When other characters speak;
However, the visual hallucinations
Do not always occur
Along with the audio to explain this,
Not that the accompanying visuals are a better state,
For they occur on top of actual reality;
Imagine even trying to walk.

I wouldn't put too much stock
In this kind of pseudo science
That says that this terrible state
Is a kind of evolution of mankind.

Schizophrenia is one of the most debilitating
And devastating mental states that there can be.

BEFORE

Before more surmising,
What are we to make of our experiences,
Whence and wherever they spring,
For here we are, receiving them
Into awareness/consciousness?

Which way to go?
Angst? Or acceptance?

Are we free of strings?
Or are we puppets?
Do we live on good fortunes's credit,
Our life but a borrowed debit repaid at death?

We are forced to choose,
The right choice, if any, unknown,
Using only the mind, perhaps,
But the mind/brain's answers are arbitrary.

Only conscious awareness
Is an indubitable truth
And it only receives thoughts
As experienced,
Not knowing if they be true or false.

We come into this universe,
Willy-nilly, not knowing,
Our lives given to us to live,
Willy-nilly flowing.

Yet, we are here in some way, no doubt,
Be it real or a very good imitation
That cannot be told apart from the real.
So, maybe, then, just be, as is?

I (Awareness—the screen)
Feel (witness in consciousness)
Happy, or whatever
(An experience that has arisen from beneath).

Yes, that's it,
So no more can be assumed,
Figured out, stated, derived, analyzed
And turned into causes,
How, why, when, and where
What has arisen, using internals alone,
Except that it is and it is now.

SERIOUSNESS

Tsk, tsk, all dross and grouching
And no lightness or gold dust;
Walk light as the wind instead
Of the heavy trodding foot
And feel the fever of life.

"I see everything," the mirror noted.

"Yes, it was LabelWench,
The lone female dog sledder of the Yukon,
Gracefully striding away into the ice fog.

She can only speak the truth—
For that talent,
Her appearance is both loved and feared.

The most humble Graybeard appears
From the mists of Australia,
Bringing the light of reason
Into the dusty haze of
The influenza death thread
Wherein humans are recommended
To fall like ants
For lack of vaccine,
As with the Indians of America
And the natives of the Pacific.

Shall we return to the days of polio?
I was vaccinated. My friend wasn't.
She got it. I didn't.

My discussion on your discussion above
Is that a real debate includes both sides.
When a contrary opinion about the value of vaccine
Is met with hints of quotes of the nature
Of 'stupidity and ignorance',
Plus what was already removed in post #749
About what the term 'mental illness' really means,
And more such already
Perhaps moderated out and yet to be,
Such as the posts about vaccine being useful
Are not at all serious, then it is not a debate
But a dictatorship or just a lecture to be taken in.

Even a spirited debate is wished for.
Note the word 'debate'. This mean pro and con.

Saying you have not even 3 minutes to answer
Is not conducive to debate.

Good luck, but this thread needed
An injection of vaccine for attitude adjustment,
As it was going under a dark cloud.

Even more now that your sources
Are questionable, Mikal,
Such a position of 'my way of the highway'
Is all the more untenable,
Not that it ever was any less.

No one will want to stay here
If nothing is forthcoming
But that you are right and others are foolish.

I will not be on a fluffy cloud,
But an active and serious contributor to ToeQuest
Who can also brighten the day in other ways, too.

Robert has faith in his moderator Graybeard
And I can't thank him enough for jumping in here
With his very astute observations.

Moderators sanitize by removing offensive
Or unrelated posts or those telling people
Where to go, even in a fluffy way.
Uh, oh, spell check almost put 'satanize'!

Let the viewers see, for once, as most is still here,
Instead of all being so quickly vaporized
And swept under the rug.

Mikal was unable to face
An overwhelming consensus of evidence
That her demeanor was condescending and insulting.

Graybeard removed some of it already,
But there is enough left to sink a haughty ship.

The current topic selected by the thread owner
Was the question of the usefulness of vaccine.

Graybeard and I were for the value of using vaccine.
Mikal posted for the fiat of that all opposed
Were ignorant and stupid
And didn't deserve a response
And so forth as can be easily and readily seen;

She believes she is an expert
And thus there could not possibly be any contrary opinion.
Perhaps this was what Everyman was trying to say once.

The day we enable anything
But discussion of a topic
Would be the day ToeQuest
Goes against its own intent.

The use of vaccines is debatable,
For example, as this is what ToeQuest is for.

I happen to see a value to them, as do some others.
Some don't. That makes discussion.

My position does not constitute a plot
Or a conspiracy against
a person believing the contrary;
One can see from other threads
That I am indeed pro-medicine.

When moderators modify or remove posts,
As they did upon my complaint recently,
And may do more of,
It is because they deem it
As nonconducive to the flow.

This could as simple as it being off topic or as intense
As it being an insult. They makka da rules, not us.

I first draw attention to straying onto the person instead
Of the idea in case it wasn't meant as such
Or if it was, then see if one wishes to work it out.

When those fail,
Plus any failure of sociability and humor,
And I observe several more insults
And characterizations following,
I report, even a whole series
If more than just one or several.

**THE MOMENTOUS COPING
OF THE MOMENT OF NOW**

History has come to an end, for this
Is where it meets the present bliss
Of now, about which tale on its axis
Swings the past and the future, a near miss.

NEGLECT

As for "those who could not answer"
And friendships being a-kin
To the ideas debated,
It is perhaps that those who embrace
The invisible myths become those ideas,
Almost literally, and so they see the debate
Turning into a personal thing.

It is not useful to go into
Any "bad" descriptions
Of personalities,
As some are ever wont to do,
Instead of the ideas,
In these cases
Of facing the contrary
(The root of evil drowned
Being their concept
Of what is good and right).

I can only assume that
Non responses to post content
Fall into some area
Of "neglect" for some,
And, at the least,
Shows some of the tendencies
Of human nature
To focus without interruption,
For their very own survival is at stake
As being the same as their ideas.

It is common enough
On ToeQuest to fully note.

ABOUT NOTHING BEING LIKE EVERYTHING...

It would seem that the information content
Of Everything would amount to zero,
For there is no wrong or right arrangement of it,
Even among those paths from Everything
That go further in time and beget life and all that.

THE EVOLUTION REVOLUTION

The Darwinian evolution revolution
Indeed overthrew the notion
Of the fixity of the species,
Science once again
Conflicting with the Creator
By showing that Man
Is neither the purpose of creation
Nor its end point.

In the face
Of the random winnowing
Of nature and time,
Purposefulness becomes a lost wish,
All the lower animals escaping
The hierarchy of inferiority to man,
Who, as we now know,
Is an animal as well.

Man emerges
Clumsily and bloodily
By 'chance'
From brutish animals,
And, ultimately, from slime.

Darwin had discovered
The truth of evolution,
Confirmed as fact since then.

Due to the prevailing [false]
Ideas of the time,
As well as his theology training,
He was hesitant to reveal his findings,
But did so, they being confirmed
By the geological strata
And now even more so
By observing evolution
In 40,000 generations of bacteria,
Although they freely pass DNA around,
And, more importantly,
In our own DNA record
That matches the fossils.

Natural Selection remains
The best theory
For the means of evolution,
The actual genetics of DNA
Arriving in the 1950's.

Could there be more ways?

Evolution by natural selection
Was a revelation whose time had come.

There is a wealth of information supporting it
And a dearth of information denying it.

The creationist argument that a hurricane
Can't cause a 747 jet to be assembled
From a warehouse of parts
Shows the great lengths
Of their attempts to deny,
That being about their 'best' argument.

If they can really supply
Scientific details against,
In order to make a solid case,
Then I don't see why
They would hesitate to do so.

REALLY?

In the 'early' state of an infinity
Of universes brewing,
Or at least all possible universes—
All these existing only within
The quantum state of possibility—
A human-type mind evolves in one of them.

Upon observation of this universe by that mind,
The multiverse of all possibilities
Collapses into the state of our one universe.
It is only this universe that evolves the mind,
The only universe that continues on to exist as real.

So, When we then look back,
We can expect to note
The rarer than rare happenings
And the extremely fortunate
Chain of events
That led to it.

ALIEN ABDUCTIONS

It seems that some may have
Such a wide open mind
That emotions such as hope and despair
Can trump the evidence of the senses,
Especially in times like these,
When the world seems to
Be spinning out of control,
When there is a surge of belief in astrology,
ESP, and other paranormal phenomena,
Spurred in part by a yearning
To feel a sense of control.

The mind looks a bit too hard for explanations,
Especially for those that let one be connected
To a larger reality, a comfort bolstered
By the thought of angels
Who are watching over one.

Yet none of this is ever so clear cut
That one can ever see it straight out.

Why is it all so invisible?
Or, to say the same thing,
"Inter dimensional"?

Because it is not really there.

Science, as it explains more
And more each day,
Actually thus pushes people
To seek even stranger sources
Of mystery and wonder,
Such as being abducted by aliens.

What makes abductees stand out
Is an inability to think scientifically.

They are often asked if they understand
That sleep paralysis,
In which waking up during a dream
Causes the dream to leak into consciousness
Even while one remains immobilized,
Can produce the weird visions and hopelessness
That abductees describe.

They say that they do
But that it doesn't apply to them.

It's really more that they didn't know
What was happening
And so they settled on abduction
As the most plausible explanation,
It being the simplest for such episodes
As a pain in the groin,
This being the stealing of sperm
By shadowy figures of ETs.

But are aliens really the most likely reason
For bad dreams, nosebleeds, and bruises?

As for other paranormal events,
It's not too far a step to see or hear
What they are thinking intensely
About through mental imagery.

DICE GAME

We've already proved that dice rolled,
With the no-beginning coming
From the eternal causeless realm
And that the mass density of the universe
Sums to the near 'nothing'
Of the quantum tunneling—
For which the proof matches
Out though experiment,
That whatever had no cause
Could thus have no design
And so there had to be
An indeterminate chaos, as seen.

So, since God cannot be,
There is nothing that He did;
That's the beauty of the proof—
It cut Him off at the source.

Then, just for fun,
And to give every chance,
We looked everywhere else,
Finding only the natural
And no supernatural.

REALITY'S DOOR STEP

John 'Somebody'—
His name rings a Bell,
A short Irishman with red hair,
Proposed a theorem that
He thought would show
Hidden variables in quantum theory;

Instead, the experiments that followed
Showed that the wave function
Of probability/possibility
Was the fundamental description
Of the quantum.

There is Everything real. (positive)
There is Nothing real. (negative)
There is Possibility. (neutral)

All must be true at once,
Especially here,
Since the information content
Of Everything,
Such as every possible universe,
Is zero,
As is Nothing's information content.

Possibility either generates the real, sometimes,
Or generates the not real,
Nothing, sometimes.

Possibility,
Being neither some real thing
Nor nothing,
Of course,
Has no laws, no forms,
Nor is anything definite,
For all is/was open
To any and all possibility.

It's the only condition
That needs nothing before it.

No more turtles standing on turtles
All the way down;
The basis was here 'forever';
Any other forever stuff
Would have needed definition.

Also, stuff having been around forever
Could also not be so,
Since forevers never complete.

Our universe full of lucky coincidences;
It was just one of a zillion paths
Of Possibility's Everything.

It is neither right nor wrong,
But it worked,
As may have some others,
But a lot of universes flopped
And went nowhere
Or were inert to begin with.

MY 3RD TOE,

Beyond that TOE of Possibility
And the one of mass-energy
Having just been around forever,
Is one that I only use for science fiction;

That the
Eternal Infinite radiates light through
A spinning complex encrypted
DNA-like code matrix,
Using information or energy
To create the CMBR
(Cosmic Microwave Background Radiation),
Which is an antenna that broadcasts waves
Of holographic interference patterns
Of virtual reality,
Even our own DNA and brain,
All tuned in by us somehow,
Or via direct access,
And is thus presented
To our consciousness as reality;

Stated more shortly as...

The Infinite radiates through a DNA matrix,
Using Information or Energy to create
The Cosmic Background antenna which broadcasts
Interference patterns of virtual reality.

EXPLAINING TIME'S DIRECTION, THE UNIVERSE, SLEEP'S PURPOSE, AND CONSCIOUSNESS

Possibility, a brute force type of 'Mind'
With all that's possible therein superimposed,
Was/Is the forge of creation,
And its echoes are our universe—
At some stage,
Whether in the real
Or in the wave function,
Having had
Great potential to be,
To continue onward.

But, what kind of observation
Collapses the wave function
To form the real from the possible?

It can only be consciousness,
For consciousness is
All the reality we can know,
Although it could be gravity
Or its own interaction.

The wave collapse is a one-way street
And so that then must be time's arrow.

There is no going back[wards],
Although it's remotely possible.

Observation creates the instantiation
Of matter that then persists.

The colors, the words written here—
They are consciousness in operation.

It is in the trees and the houses—
It is the feel of those things...

The odours, the mountains beyond—
They are all consciousness.

Our only portal is consciousness.

Bell's theorem shows that
Nothing "out there"
Has an entirely independent existence.

We can never see "out there";
We only ever see the inside of our brain!
It only seems literally out there
Since the brain projects it
As being out there.

Colors, form, textures, odours,
Light and everything exist
Only in the brain as such.

The brain is amazing
And can sort a 100 million bits
Of information in an instant.

Consciousness must play an active role
In the functioning of the brain,
For both evolved together.

They are intertwined in a process.

It's also a way for one to actionize
Without actually moving.

Consciousness,
Perhaps mechanically fundamental
Since it is so unlike all else,
Must still yet tie into
The regular physical, and vice-versa.

How is consciousness mediated
In order to interact
With the physical world?

It must be at the synaptic cleft
That what we call 'mind' qualitatively
Meets the brain proper.

Here is where the neuron fires or not;
Here is where a neuron meets other neurons;
This must be where data turns into thought.
Whatever triggers these switches
Produces thought in consciousness.

But how does consciousness operate?

The nonlocal properties
Of quantum mechanics
Might play a role,
For, any of the regular physical forces

Such as electromagnetic, weak, and strong
Would be too disruptive
And/or not have enough reach.

Consciousness may be
Of a quantum biological nature;
So, then, it would be tied
To the quantum mechanical process.

The electrons reach everywhere,
Tunneling and hopping about,
Then take the most useful path,
Perhaps globalizing the results
Of the synapses
By selecting a thought or an action
From the superimposed scenarios
Of the action's consequences.

This is the will selecting a course.

The long-range electron tunneling
That connects synaptic firings
Throughout the brain
Into a self-sustaining pattern
Is consciousness.

This contact, then,
Between synapses and electrons,
Reaching across the whole space of the brain,
Turns on the light of knowing.

Consciousness occurs
Above a certain synaptic firing rate,
While sleep ensues below the limit.

Sleep is to allow the ground state
To be refreshed;
Consciousness has to pause
For restoration
When it gets worn out.

This was so important
That evolution retained sleep,
A time of grave danger to any species.

Melanin
(What is it doing inside the head?)
Absorbs many of the electrons,
But not all of them.

Sleep accounts for the rest,
Allowing molecules
To relax their excited states.

While time's arrow is always pointing forward,
Since the quantum wave function collapse
Is a one-way street,
There is the 'not time' notion
Of no time,
That is, nothing happening—
No movement of any definites
And therefore no change and no time.

Time, said someone,
Is Nature's way of preventing
Everything from happening at once.

Yet, in the neutral position
Of Pro Theory,
There is the timeless,
Which is also in the realm
Of the Possibility TOE,
That is neither time nor 'no time',
But an all-at-onceness.

THE WAVES OF THE ANCIENT SWELLS
OF UNFORGETTABLE TIDES

Time hurled a million waves of displacement
At Memory, yet it was still there,
Being outside of time.

Time, now gray with age, hurled its changes
'Gainst Memory's rock, time and time again,
Entropic seas denuding the sands.

Reminiscence weathered,
But could ne'er wither;
For, in those mists of time,
Yesteryear yet appeared.

IN NOT TIME

The probable emerged from the possible,
It's future already past in an 'instant',
For everything has already 'happened',
Although we're just conscious about it now.

FLUID DYNAMICS

Two hints of a wisp of a breeze,
Each the other way going,
Passed near to each other,
At the equator, this time swirling
And continuing to spiral,
Ever more and more,
Eventually giving birth to the beast
That was Gustov,
The blob that ate Santo Domingo
And spit it out,
After raking Jamaica lengthwise,
End to end.

The whirlpool, a category 4 hurricane,
Drew energy from the water,
Then grew larger in mass,
Which then produced more energy—
A feedback system out of control.

It clipped Cuba's feet
And drowned the outer Keys,
Heading now for the warm waters
Of the gulf to refuel and perhaps growing
Into a humongous monster of 5 or 6.

Gulf oil platforms had been emptied,
And the roads of the coastal towns
Were all one-way out, both lanes—
No return possible and none desired.

The new New Orleans
Awaited the hand of fate,
Tempting doom
To chance its way once again.

So, too, perhaps,
Our universe of whirling forms
Began as such,
As two oppositely flowing streams
Of fundamental energy-substance meeting,
And then whirling/twirling
Into spinning nucleons
That threw off photons
At their own spinning speed of light;

Then larger forms whirlpooled into stars,
And galaxies turning,
The voracious beasts of black holes

At the center devouring all those
Who entered the Gates of Hell—
All hope having been abandoned,
A one-way street to oblivion.

CIA/DIA/NINJA

The DIA operatives had been listening
To the General's rapt descriptions
Of the prelude to his major adventures
That led him to become
Their commander and GrandMaster.

They had reconvened in Niihau, Hawaii,
His home office that he'd often left
So as to be active in the field—
The place where he was meant to be,
The place where he was ready to operate from—
The place that so much high tech now allowed.

They could hardly believe their eyes
When they saw the place.

No one was in a rush to hear all,
For they ever listened intently,
And savored all the details
As they slowly but grandly came forth
Via the nearly lost art of story telling.

So, they were not impatient,
And could wait,
But they could feel that
Some things writ large were coming,
That stuff behind the scenes
That "never happened",
It being just beyond the horizon
Of the recent and ever building stories
By this lone General GrandMaster
Of the Ninja Empire
That protected the world's innocents
As best they could
With their deep determination
And ever-present goodness.

And so the General, alias Rascal,
Magic Dragon, and GrandMaster
Unveiled more tales, narratives,
Anecdotes, reports, accounts, and sagas.

WHEN I HAD MY NDE DID I GET
ENTANGLED WITH THE GRAVITATIONAL WAVE?

I can only speak for OBE's,
But the 'out of body' part is similar for NDE's;
This could fool anyone.

I was fooled the first time;
I saw myself from above, clear as could be:
'Out of body' and seeming
To look down at my actual body.

In a subsequent episode I was lucky enough
To note my real-looking 'dream' arm,
Although not fully opaque,
Diverging from my real and unmoving arm.

I was in sleep paralysis;
This is when one is partly really awake,
But cannot move.

When we have an normal imagination
Or a memory of a scene, we often see it from above.

Think of Graybeard and his girlfriend
Laying on the beach of the Gold Coast
Instead of researching ant food tastes;
You see them from above.

It's not a vision that comes in an OBE,
But of any sense;
Once I kept a dream song playing
For 10-15 seconds after I awoke—
It was playing only on the mind-brain 'radio'.

One may have an urge to reply
That the OBE/NDE seemed really real,
That it was surely not a vision of imagination, etc.,
But just think, first, about how it is always
And only the mind-brain that puts a face on reality.

I cannot stress enough that we only ever see
The insides of our brain—the reality show
Whether awake or asleep.

Sure, objects seem to be 'out there',
But they are actually projected,
In the mind-brain 'eye',
Out back whence their light waves originated.

There is no color, form, texture, etc. out there.
There is not even light as we think of it,
But 'light' made inside of our dark heads.

When one is 'floating' above one's body,
It is not that Gravity's laws have been repealed,
Nor is one in another dimension,
But just in the mind, as always.

Plus, in an NDE, a Hindu, say,
Does not see the same religious visions
That a Christian does,
Further showing that all
Is of the individual's mind-brain.

To boot, a full blown OBE experience
Can be induced chemically,
But no one really wants to have an NDE induced.

In an NDE, one is in danger of death
And so the brain is certainly not in a normal state,
Perhaps even being drained of oxygen and nutrients.

Seeing a light at the end of a tunnel
Is a result of how the visual cortex
Works in this state.

We normally only see clearly only
At about the size of a deck of cards
Held at arm's length
(Try looking just a little away
And the clarity goes way down)—
This is the center of the tunnel
Which is caused by neuronal stripes.

In summary, one sleeps, meditates, or collapses,
Then has an OBE/NDE, and then gets up again,
All going rather back to normal.

*Have studied sleep paralysis and compared it
To my own OBE... it is the state preceding
The ability to experience the OBE state.*

Yes, this is indeed the key to the apparition.

*None of this has anything to do with imagination.
Which is experienced in your mind
And is the mental effort to make mental images...*

This is, of course, not an effort to try to imagine;
It is the imagination just going on its own,
As in a dream;
As many would swear to a dream being real,
So would they for an OBE.
It is not just from books that OBEs
Are well known for what they are,
But also from the laboratory.

NDE: I am not really dying to go down the tunnel...

If you fill in the overall perspective
With inappropriate details
Because that is what you believe,
Then the overall view is automatically blurred.

This relates to a person's capacity
To become absorbed in his experience.
For example, someone who easily
Becomes immersed in nature,
Art or a good book or film or a computer game,
To the exclusion of the outside world,
Would be one who scored highly
On the scale of 'Absorption.'

Irwin expected OBEers
To be higher on this measure
And that is what he found.
His OBEers seemed to be better than average
At becoming involved in their experiences.

Yes, this is what I read, too,
Which is why some OBEers
Are more prone to believe
That the experience was
Actually real instead of just a vision.

When one is half asleep but half awake,
Or even half dead or half alive,
One is in a mixed state of both.

We surely know we had a night dream
After we wake up,
Even if it seemed so real at the time;
But, when we totally know that we are awake,
Often just partly, then we surely believe all
And would swear that
The visions are of the real world.

Yet, as in my OBE music
That kept on playing after I awoke
Or in the OBE visions while awake
But in sleep paralysis,
The visions are surely being manufactured
By the imagination of the dream world,
Yet playing in the awake world,
Sometimes even on top of it
And sometimes even making a blend
Of the two worlds,
As when I saw my dream arm
Diverging from my real arm...

This is not imagination... no imagination here!!!!

As you can see from my example,
The awake and vivid dream imaginations
Of an OBE/NDE are indeed a very powerful
And often convincing occurrence.

It got me the first time
And I really relished
The experience of floating around,
Even fiddling with some things
(But, later, they were seen not to be moved).

It is really not likely that one left the body
And went thousands of miles or even 5 feet.

ALL IS ILLUSION?

I wouldn't use the word "illusion"
As that is suggestive of a delusion,
Misapprehension, misconception,
False impression; fantasy, fancy,
Dream, chimera, or deception.

How about a "model" or "interpretation",
For that shows there is something
Behind it to be modeled or interpreted
Instead of the nothingness of an illusion?

Anyway, enjoy—it's the Lifetime channel.

ERRORS

A type I error,
Or a false positive,
As believing something is real
When it is not.

A type II error,
Or a false negative,
Is not believing something is real
When it is.

Believers in UFOs, alien abductions,
ESP, and psychic phenomena
Have committed a Type 1 Error in thinking:
They are believing a falsehood.

... It's not that these folks
Are ignorant or uninformed;
They are intelligent but misinformed.
Their thinking has gone wrong.

In the search for truth
It could be that we ignore
Evidence of the truth,
By 'neglect',
In order not to be 'duped',
Or because it is not what we wish.

So, we believe a falsehood.
This is a type-1 error.

To compound this,
Sometimes we go on to believe
The opposite, regardless,
A type-2 error—and get duped anyway.

The terms Type I error (false positive)
And type II error (false negative)
Are used to describe possible errors
Made in a statistical decision process.

Type I: reject the null-hypothesis
When the null-hypothesis is true, and
Type II: fail to reject the null-hypothesis
When the null-hypothesis is false

We must be able to reduce the chance
Of rejecting a true hypothesis
To as low a value as desired;

*And the test must be so devised
That it will reject the hypothesis tested
When it is likely to be false"*

*A type I error, or a false positive,
Is believing something is real when it is not.*

— Gravity can be suspended.

— One can communicate with other dimensions.

— Was not in sleep paralysis.

What is more likely, a miracle or not?

— Condensed from Michael Shermer,
Why People Believe Weird Things, Introduction

THE BIBLE

While it's interesting to read
Or translate the Bible
For reasons of learning
How humans thought in those days,
Its accounts fail to match
What has been found by science,
For example, space not being a firmament
And the Earth not being a center of anything,
Plus, in general, that the revelations
Are not revealing at all,
Being that the predictions
Never came to pass,
But for some written
After the facts of the past.

Many have seized on Biblical 'truths'
Such as that slavery is permissible, and so forth.

The Bible's One is basically an amalgamation
Of the many old Jewish Gods of legend.

Now that God has published
The #1 best-selling book,
I heard a rumor
That He may even
Enter the modern media
And come out with a movie.

MAGICAL HAPPENINGS

What secrets of life and death
Lay buried in the sands?

What inaccessible truths
Protect themselves by their own magic?

The old Rascal lit up a cigar,
And the stories unfolded
In the haze of this pipe dream...

"Do tell what else
Was in that Great Pyramid, Fredrick,"
The General suggested.

"There were 4000 year-old iron weapons
That did not rust,
Looking as new as the day they were forged.

"I held glass that bent without breaking.

"I drank from a vase
That poured water without end;
I filled an entire tub from it
And bathed away all my dirt and dust.

"A compass needle went around
And never stopped.

"I ate a cake but I still had it.

"I saw the starry skies
Through solid rock walls.

"I entered a room that had no door.

"There was light within the room
But no flame or openings.

"I looked into a grain of sand
And saw eternity."

Fredrick paused, recalling.

"Outside, I saw the Sphinx.
Its glance was fixed on something else.
It was the glance of a being
Who thinks in centuries and millenniums.

I did not exist and could not exist for it,
For it was the face of eternity."

No one spoke.

The General rose.
"Next, after an hour break,
During which you might go out
To see the scenery,
We may hear some about a long trek
From an escape from a Soviet prison
Through the mountains
And across some ink-black rivers."

Questor and Top Secret
Headed down one
Of the many paths of Niihau,
Its secrets ever shrouded in mist from above
And all around from the other islands;
But, here they were,
Within the Forbidden Paradise...

"Come back, friends,"
Said the General,
"To hear of the dark,
The light, and the never."

"We are here, being ever."

"There are books unwritten and never told."

"We can listen until we get old."

"By what muted shore of the dark river
Did its strand call me forth?"

"We're sure that we'll never hear worse."

"By what far edge of furrowed forest
Didst the Motherland seek my name?"

"Oh, Dragon, through what hazy depth
Of gloom hast thou tread and threadest?"

"Gather thee round and you shall knowest."

GOD DISCUSSIONS

To discuss a particular idea of God
No more impugns character,
Even indirectly, than, say,
Discussing pro and con
Of whether there is an actual reality of space
Or if it is just an empty place.

Otherwise, all sides of a debate
Could be charged with impugning
The actual characters of the debaters;
Lame reason to try to stifle debate.

Ironically, in reply to this lame claim,
ToeQuest is here exactly for such debate!

The 'evil' is trying to protect
The [flawed] 'concept of good'
By undermining the character of those contrary,
A common ploy used before on TQ and in life.

Kicking the player does nothing to support one's ideas.
In fact, it probably even dissuades and distances
Some members from taking the idea seriously
And/or wishing to discuss it further.

God is not just an idea—
But an an idea about an invisible;
So, it is a theory.

We've explained through natural selection
How humans naturally came
To believe in good and evil spirits,
All of which later on were shown to be wrong.

The Gods were not on mountain tops or the moon;
Physical ills were not caused by evil spirits
But by bacteria and viruses.
Mental ills, called sins,
Were not of the Evil Spirit, the Devil,
But of brain chemistry,
Upbringing, and human nature.

Nor was the Earth flat or the center of anything.

The Jewish even dispute the divinity of Jesus,
They being there at the the time,
He even being one of their own.

Anyway, from there—
The wish of an idea to be true—
Religion makes a leap to that it is a truth.

This is an error—to preach a theory as a truth.

Until 5th grade, I was one of the little ones
Preached to by this merciless indoctrination
By the Catholic Church.

From the invisible,
A whole further structure
Was built upon more assumptions,
Umpteen levels high.

Religion retarded science,
Burning and pursuing the scientists;
The Catholic religion (and others)
Even attempted to squash other religions
With crusades of killing,
Even those with relatively minor differences.

We would just leave it at that, if it were harmless,
But Religion still ever attempts stifle inquiry,
For it there is no further questioning of its answer
That 'God did it."

So, there are clashes about
What sciences should be taught in schools,
As if the known can't be taught,
In favor of the unknown,
Not to mention that the Church
Still lobbies Congress
Against scientific progress.

To boot, God, as known in the testament of old,
Is not even a good role model,
But is even a menace of
An emotional vengeful Person.

Good riddance to that Guy whose traits
Are not good to follow,
Joining the forgotten lot
Of those hundreds of gods come before.

That insanity
Couched in words like 'mysterious ways'
Makes the Guy look very much
To have had a human nature imposed on Him.

So, these are thoughts on the idea
Of [the recent] Imaginary Friend—
God, and the actual instances of religion's beliefs;
There is nothing here about personal character.

THE ANSWER THAT ONLY QUESTIONS

If you require a basis,
Of God,
For the Eternal uncaused FS,
That 'simple' stuff,
Then you must all the more
Require a basis for
This Ultimate Complexity
Of a Creator
Who does planning, thinking,
Moving, and designing—
Or else drop the requirement
For the initial and lessor basis.

One cannot have a cake and eat it, too,
Being inconsistent by doing
An analysis half-and-half,
Shoving away the answer,
For then a stop sign goes up
That is not an answer
But just a larger question.

THE PYRAMID

The Electric and the Magnetic forces
Transition into each other,
As a self-renewing electromagnetic wave
That can go on at light speed toward 'forever'.

These points are like as East and West
Blending into each other on the globe.

The opposing forces of Weak and Strong
Are of changeability vs. stability—
They are as different as North and South

Gravity arises from all the forces together.

The Strong and Weak opposition
Perhaps shows the separation of the forces
Once thought to be able to be unified.

GETTING POINTS ACROSS

Luckily, English has many synonyms
To help with the preciseness
Of our attempted conciseness,
But, still, down deep, sometimes,
There may surface wordless thoughts
Which can only translate as 'ugh'
Or some feeling that some notion
Feels great or true,
Again in some vague
And unworded speechless way;

So, then we must interpret
And even sometimes invent
The rest of the story as best we can.

Even in poetry,
In which we force ourselves
To be precise,
To get the message across
Within the constraints of the form,
We might only get across
A general interpretation
Of the "soul's" feelings.

In fact, this presentation of all
That is possibly not so readily apprehended
As truth is the very purpose of poetry.

Poets translate the "soul' s" thoughts
And feelings into words
And try to send them forth finely dressed
For others to read.

Is this the end of it—
The poetic words?

Did we really write a total poem?

No, not necessarily,
For the poem must induce the reader
To then translate the words back
To the depths of the soul feelings
Of its sometimes speechless mode.

It is only then that one can say
That one has written a successful poem,
That is, only if the reader's soul
Is responsive to it.

CONSCIOUSNESS VS. CONSCIENCE

Consciousness is our only portal
To what's out there
And what's in here, in the brain,

Therein being a representation of it.

We only ever see the insides of our brains;
Amazingly even light
(Inside the 'dark' brain)
Is a representation.

We fiddle with what's outside
And so then we know that
The representation is faithful.

I would guess the representation
Is an even better and improved face on things,
It being lot easier to deal with
Than having waves and fields or whatever
Somehow impinging more directly
(What a mess of non-sense that would be).

Only about 5% of what
The brain's analysis does
Actually surfaces to us,
After 300 ms., into another,
Higher, focus of the brain
Called the mind,
But just a few things arrive at a time.

Consciousness,
Being another part of the brain,
Or some global binding,
Is the witness to
What is currently on the 'mind';
We call it 'I',
Just as in normal English.

That simple case is not the end of thought,
For we may ruminate,
Either rejecting some simpleton thoughts
Or letting them feed back globally
For more analysis by other brain areas.

The end result is always of what
We have become during our life.

Many activities,
Such as those already learned,
Like driving a car,
Don't need much attendance
Of the awareness of consciousness;
However, learning to drive
Needs full consciousness.

Have to have food right now!

There's no source nearby;
How about the next town?

Of all our thousands of thoughts
And senses a day,
Only a few might give our conscience
Some pause to decide
The right path from the wrong,
Or at least if it is a brand new instance
Needing review,
One that we don't already know.

THE 3RD CHIMP

Darwin told us how natural selection
Explained the mysteries of evolution
And the variety of life.

The continuum extended
From animals to us.
We were no longer special,
Differing from chimps by not so much DNA;
The discovery of genetics later collaborated all.

FEAR CAN BE TURNED INSIDE OUT—
INTO EXCITEMENT!

Say, for example,
One has a fear to speak out in public;
Yet, looking inside out,
It is but the excitement
Of having many listeners
Attending to what you say,
Plus, they are all naked!

THE MEDIUM WITHIN GROUND ZERO...

Here, there and everywhere,
[no particular time, place, or form]
Unconstrained
[all was open: lawless]
As chaos reigned
[the ultimate disorder]

Supreme, until realizing a litany,
Light did shine
And drew the line
Back and forth
From brightness to darkness.
[a fantastic happening,
leading to the spewing of stuff]

Blinking back starkness
Till, fully awake,
Excitement fluxed and glowed,
To ultimately explode!
[inflation]

With Awareness Comes Expression

Riding a Planck, wave-surfing an ink-dark void,
[the uncertain Plank realm]

Exclamation employed Punctuation;
Swift violent journey from obscurity
Created the purity of aether's roiling,
Foam of swirls and oscillations,
[like a roiling sea]

The enigmatic implications
Still weaving the crest at speed;
Much unfinished
And yet, undiminished.
[simplicity ever begets more complexity,
but, it takes time]

Loop-de-loop In Plasmordial Soup

A glance behind,
As bright sparks ignited the wake
To persist and propagate,
Incendiary brilliance flared;
The multiple creations
Burgeoned inflations
Bloomed to attract, collide,

Entangle and coalesce,
Populating the emptiness;
[clouds of stuff, collisions,
Or a parallel to life's soup on Earth
If we look ahead]

Then, evidence the immense,
Spectacular vortices;
Synaptic relay to cortices.
[stuff ever swirls; stars or more life on Earth]

Outward and onward, accumulating information
Combined with inflation
And intent, to impart long journey's record:
[more complexity in space,
making more complex atoms and molecules,
or the again the parallel
of now much higher Earth life]

Be of one accord,
The abstract spinning symbols
Of rock attracted,
As light diffracted;
Cohesion and order
Of formations spherical,
Now waxed lyrical.
[bless you, stars,
Their emanations, and life]

And so, in becoming attuned
To the movement;
Room for improvement,
Leaving behind multitude dull,
Failed attempts;
[other bubbles that fell flat,
either of other universes
or areas of our universe
or planets not making it]

Then, unlikely event!
Arriving at the one place
Of brilliant perfection;
In every connection
Frequency intensifies
To seamless harmony,
Ordering the hegemony.
[us or our planet]

Architectural Vernacular and Instant Teleonomy

Finally, having discovered great capacity,
Beyond mere tenacity,
Behold!
Every conception of life in instantaneity,
Naught in extraneity;,
[everything was/is necessary, all connected]

All purpose is focussed
Upon this manifestation
Of ultimate creation.
The reflection of thought awoke,
Arose to share;
[consciousness]

Drew first morning air...
[life]

TO ESCAPE THE PARADOX OF

Nothing ever becoming from nothing,
And of Real stuff always having existed
In a definite form and amount and more
Without ever having been defined to this state
By its having been around forever,
Which is a problem in itself since
Eternity could not have already passed and completed,
We surmise that the only "thing"
That needs nothing before it is Possibility.

Possibility is even the naturally extended
Next dimension of all things in superposition,
Such as all possible universes.

All was wide open "then",
For there were not any physical laws,
Nor any form of substance,
Nor substance moving
And causing what we measure as time.

A not-quite but perhaps close analogy
Is the quantum realm
In which a "particle"
Is everywhere but nowhere.

HIGH SPEED REPLAY

The journey of Human's evolution
From the raw material was recorded
And is remembered by RNA/DNA,
It then (re)playing out 'quickly'
Through the nine months in the womb
And right on into the beginnings
Of first consciousness
And then further
To self-reflection and growth.

THE SOUL

About the supposed, um,
Immortal spirit soul
Conjectured type entity...
Just a small point...

A point rather minuscule, microscopic,
On a nanoscale, infinitesimal, little,
Mini, diminutive, miniature,
Scaled down, Lilliputian,
And teensy-weensy, itsy-bitsy, eensy,
Eeensy-weensy, little-bitty, bite-sized,
Pint-sized, and wee... um...

This angelic vapor of a spirit soul
Is not seen and so it is
That someone made it up,
Perhaps out of a wish for it to be true.

They then went on to build
A whole further 'structure'
Upon this invisibility,
Such that it goes
To Heaven or Purgatory or Hell.

PRICELESS

Antimatter is considered
The most expensive material on Earth;
A commonly quoted figure
Is that it costs
$1.75 quadrillion dollars per ounce.

EVOLUTION NEVER WAS?

If evolutionists understood ancient history
The only evolution they would be discussing
Would be in the nature of psycho-spiritual evolution
And the expansion of human consciousness
To a level of intelligence capable of creating a world
That we have been entrusted with....

Smoot wrote of the obvious parallel that exists
Between the big bang and the Christian teaching
Of creation from nothing.

Indeed, a parallel;
And this is why science
And religion appear to converge,
But one view wishes for perfect order,
The other shows complete chaos—poles apart yet.

However, as scientists looked at the odds,
They soon realized that a random explosion
Like the big bang could never have led
To a universe compatible with life.

They most certainly do not.
This is easily verified.
Now I am suspicious of this source.

Irreducibly Complex Organs

This is old stuff;
Even the flagellan's propeller
Was shown not to be.

Behe defines these organs
And systems as irreducibly complex.

Read up more on this guy; he failed.

Materialists like Dawkins argue, however,
That it is possible to see
How the eye could
Have developed gradually,
Like Darwin theorized.

And he and others long ago went on to show this,
Along with in-between transitional 'eyes'.

However, to date, no scientist has been able
To adequately explain how unguided natural processes

Could have produced these
Irreducibly complex biological systems.

As said, really old news.
It has already been done;
Fortunately, many more events beyond
These old approaches have transpired
Over the last many years. I posted many.

"But as by this theory
Innumerable transitional forms
Must have existed why do we not find them
Embedded in countless numbers
In the crust of the earth?"
— Charles Darwin

Old news. I showed this.
Google 'transitional forms'.

So far, we have seen that
There is no fingerprint of God anywhere;
Plus, theists have not shown that
There is any fingerprint whatsoever.

It doesn't go anywhere for someone
To just hope that there might be a fingerprint
Or that there will be one.

Where is it then?
Hopes and predictions are not facts.
No moon God; no Mt. Olympus Gods;
No sun God; and now no universe God.

This is not terrible;
It's just that we search for the truth.
Truth is what can ultimately help
The human race, not myth.
The thing, though, about the theists saying
There should be fingerprints of evidence all over
Verifies the approach of
The disproof of God that finds none.

Note that only one fingerprint would show God;
One measly one, that's all, but, still, there are none.

It is really that humanity is fine-tuned
To the earth and its part of the universe;
Evolution by natural selection has shown us that.

The IDers of the past have been confined
To looking for gaps, but now the gaps close
To squeeze them out of their last refuge.

Behe, for example, just put out the same content
In a new book that was already discredited
In his previous book,
He not even including the new finds thereafter.

For one glorious mishap of his a while back,
Google Dover, Creationism, court trial (Delaware?).

Same for Ross.
It's always good to investigate before citing.
I leave it up to you,
For then you will have less qualms than if I post it.

Yes, all in the universe is
Indeed connected by all its interactions;
However, there are many outcroppings
Of forms therein.

In dealing with God, there is nothing visual,
But the whole is still wished for in the mind,
As should be the implicated details,
But only if we allow them to have a place,
For it is easy to neglect those.

Some might just remain
At this level of "all is one",
Forever placing themselves
Far away from any detail,
Even to the point of not
Being able to consider them.

Furthermore, the mind,
In the case of the imagination
Of the idea of God,
Contains no real and actual details
To guide it for any armchair analysis.

To be complete in examining a concept,
The brain must be able to think
Both scientifically and holistically,
Although these views
May not present themselves
At the exact same time;
We need to juggle them perhaps,
If the holistic is dominating.
But, just remember that we

Are already one step lost
When we even begin
To imagine invisible things.

The problem with all Designer arguments
Is their failure to account for the nature
And prior existence of the Designer.

Under what set of physical laws did God operate
Before He made the universe
And where did those laws come from?

We are left in the same Ark as before
And so we cannot just have the right brain
Exempt God from these real concerns.

If He was Intelligent to enough to engage
In large-scale cosmic engineering,
By what means did He evolve that intelligence?

How did He know that the system of natural laws
He chose before the Big Bang would inevitably lead
To thinking creatures like ourselves?

And why did it take so long
If the answer is that He is all powerful?

And if He existed before time,
Then how did he act and plan and make ready
For the genesis that was really just an experiment
(As if He doesn't really know All in advance)?
It would even take time to invent "time".

One cannot just have a great system of mind
Being responsible without allowing that all systems
Must have components beneath for their operations.

Name calling does not add proof to an idea,
But rather detracts, even sidelining the discussion
Into the whys of the insults.
ToeQuest is wise to this kind of maneuver.

FADS

Yo-yo's and hula hoops are done,
But poems still bring a lot of fun.

THE AMAZING AND UNEXPECTED
SUPERTOE RESULT
THAT ALSO DISPROVES
THE SUPERNATURAL

We have found the glorious understanding
That what ground-state gave rise to the universe
Must itself have been eternal and causeless
It being a chaotic and totally arbitrary state,
For there was never a time of its birth
It necessarily going back forever,
From which "place" it could not
Have been given any definition or intent.

So, then, there is no more regress of infinite causes
That could never be; nor any more thoughts
Of all coming from a Nothing,
Of which not anything could become.

The buck has stopped here
And we have cashed it in;
It is the currency of the quantum fluctuations,
These being a near 'nothing',
But not a total Nothing, and that is
What made all the difference
In the world and the universe.

DINNER BREAK

Anti-pasta is overtaking pasta.
Never consume both at the same time,
As you will then explode from both ends at once.

LIVING FOREVER

If we can put a better cap of the ends of DNA,
Where some junk wears away,
Then the good DNA after it won't get ripped away
After a certain number of cell divisions;
Then one can theoretically live forever.

Just be careful crossing the street
For the one zillionth time.

WISHING

In general, about speculating...

Since faith is belief in the absence of evidence
It should not be used to make any judgments
About the world or personal life.

It is also that what is eternal must be causeless
And so cannot have been defined,
Such as a complex composite mind,
For there was no point at which to do it,
Not even to mention that the parts
Of such a system of mind
Would be even more fundamental.

So, an eternal causeless God, even more so,
According to Leskey's design principles, and mine,
Would have to have had a DESIGNER for its own order.

Sometimes the mind so much wants to do it's function
(To know all),
That it speculates its way to 'truth',
Not realizing that its mere pronouncements
Just float in the thin air as an unsupported belief.

ONWARD AND OUTWARD

The 'vacuum' quivers with uncertainty.
These twitches, each one near 'nothing',
All tolled, put out a lot of of conserved energy.

Position and velocity are a complementary pair,
Each inherently uncertain.
Only together do they form the complete picture.

The same with the wave and the particle;
To see one destroys the other.

The universe was simple,
Way back when;
It is only now that its combinations
Have become complicated.

SUMMERTIME FUN;
HAVING A BALL

Everyone was out having a ball lately
By hitting it with a club, bat, or a racquet.

Graybeard shot an eagle and a birdie
And then cooked them for dinner;
SB_UK scored a wicket, whatever that means;
Arthur argued the laws out in left field
With an umpire who was always right;
LabelWench jumped her horse
Over a giant snowball;
Max rolled a bowling ball
Down his road at ten trees.
...
Austin avoided the net of evil
And tried to keep within
The white lines of goodness;
Mikal thought that a sand trap
Near a water hazard was a beach;
MJA's equal game ended in a tie;
...
Melanie scored 18 holes in one
Because it was really
The Perfect Awareness
That was playing;
TimeParticle hit a golf ball
With a baseball bat
And created a new orbiting moonlet;
Bogie shot a bogey;
ProfPat was free
To hunt for any game.

Any more?

LW, I never played polo, but I played golf,
Which I learned from playing billiards,
At least the putting part,
And so I suppose polo is really
Golf combined with riding a horse.

TimeParticle is so strong
That he hit the ball out of sight;
Melanie said that there is
Really no horse and ball;
Graybeard reached for a branch
When the horse's ears twitched.
...

Graham used a levitating magnetic horse
After the bull moose threw him off;
Austin went hoarse from too many posts;
Wick played from the 4th dimension
With a hypersphere.

Everyman was dying to find out
What happens after you die
And almost died laughing,
Nearly finding out.

NOW ONTO THE UNIVERSE...

For certain, we are of the universe,
Part of the universe,
And actually <u>are</u> the universe,
Whether before birth, during life,
Or after death (recycling).

Consciousness is seen
To be a kind of universal
In that it is identical for all,
Although its content differs,
And so it is that consciousness
Ever develops and follows along with what
A sufficiently complex biological medium
Comes up with (so far not of silicon or anything else).

So, it is consciousness that ever witnesses
The light of your brain's analysis
That derives from your particular learnings,
Memories, and associations therein, from stardust.

And, in the multiverse, if there is one,
We are but a gleam in the bubbling quantum soup.

LEARNING

False foundations eventually meet
Their destiny of crumbling away.

People spending beyond their means, crash.
Mean people run smack into the wall of life.

THE CONVERGENCE OF SCIENCE
AND THE NEW DEIST 'GOD'

Some theologians now say
That it is no longer important
Whether humans evolved or not;
However, this is not the traditional God.

Be this as it may,
Many Christian theologians
Seem to be converging
On this new view of God,
For they had to.

They are serious thinkers,
Having gone a bit beyond
Just making everything up;
It is in their interest
To seek an objective proof of God.
It's just that they couldn't, so...

It is the new deist God where

1.
God does not directly act in the universe.

2.
Homo sapiens is an accident.

3.
What happens in the universe
Is not according to a specific plan.

4.
The new deist God creates the universe,
Giving it all the elements of chance.
...
There is no reconciliation, really,
Of this new God with Christianity,
So it is hard to see how
They will make this God into one
Who should still be worshipped and prayed to,
For He would not answer. He left. He's gone.

One might note that this God
Might as well not even have been there,
For He did nothing extra
That was beyond the natural.

— 136 —

This is a God who played dice with the world;
A single toss of the dice made the universe,
Just as scientists have always thought.

GOOD OLD GOD

Poor old ageless God has been here forever,
Knowing everything, and was never born,
And so He found Himself not even being able
To take credit for Himself
Being the One and only Boss.

He wondered where He came from
But could never know,
For His earliest recollections always eluded Him.
(Note: so He couldn't know everything.)

He would have made a universe
But there was nothing to make it out of
Since He Himself infinitely used up Everything.

Plus, what would be the use,
For His designs would all be perfect,
And, if not, He would already know
The outcome if He made them faulty.

The phenomenon of reliably consistent creation
By causal intelligence lying behind it
Is philosophically and logically impossible
Without more causal intelligence lying behind it, etc.;
That is, a system of intelligent mind is a system,
Having parts beneath that are more fundamental
Than the resultant system.

Where does it end (begin)?
It cannot be with mind, for mind is composite.

It is just conjecture to say
That God's energy made our energy,
Plus now we have to wonder
What made His energy.

Instead of the answer we think we have found,
We've only created a much larger question.

So, without much else to do,
God made a square circle,
But this can't exist as much as He can't.

MY GOD

If God exists he almost certainly would have to be an alien.

True, simplicity is ever found below;
Higher complexity is found 'above',
Such as a more highly evolved alien.

We have searched high and low, near and far,
But have only ever found the natural,
And never the supernatural.

It seems apparent that life must have originated
From lifelessness to begin with,
And may do it fairly often.

Yes, god is not needed.

One must still not preach "God did it",
But, to be honest should preach
That God is only a theory;
It is unfair to lie and say that it is a truth;
For, the invisible is still the invisible,
No matter what the proposition.

If gods exist they would necessarily
Have to be technologically advanced
Far beyond we humans on Earth,
To the point that they became gods.

Yes, look to the more evolved
Or to the future for 'gods',
Not to the past or the beginning
Where simplicity reigns.

If God exists he almost certainly would not be
Restricted to any particular body, form, or gender.

Humans built Him,
With this image of the strict Father
In the image of the human family.

If God exists it seems most likely
That he has as much influence
Over the content of canonical texts
As he wants to have.

God must have made them wrong
Since there is no firmament,
No Earth being in the center,
No immutable forms, etc.

If God exists, it seems quite clear
He makes use of the evolutionary method of creation.

He would have to.

If there are things which people consider to be spiritual,
They are most likely actually physical
In ways we just can't appreciate yet.

Yes, nothing but only the natural
Has ever been found.
Extraordinary claims of souls
And such have no evidence,
Much less an extraordinary evidence;
Faith is a belief in the invisible unknown,
Even admitting that there is no basis.

TRIPLE ERROR

Spirit believers go to desperate lengths,
Such as to ignore visible facts
In favor of the suppositions of the invisible realm.

This is the extreme double error
Of the false positive and the false negative,
But, there is an even worse stance.

If it's not bad enough, that is,
To ignore visible fact
And to also state that the invisible
Is not just a theory but a fact,
The third error, then,
Reaching the ultimate limit of errors,
Is to then build a whole further structure
Upon the invisible side,
Such as God's intent and so forth,
As contradictory as much of that is.

This then sometimes even reaches
The ridiculous of states of creationists saying
That God put fake fossils in the strata to fool us.

THE POTENTIAL OF THE GROUND-STATE

The Potential or near 'nothing'
Of the quantum realm,
Is shown to be an indeterminate
Disorder of chaos by science,
With no hidden variables within.

This matches what we would also realize:
That it can't be a system, a mind, a design,
Or a designer in itself, for there was no prior time
For the eternal causeless itself to have been designed,
As there is neither anything before it
Nor any more fundamental parts to make it of
Nor anything or any way to design it,
Especially all the more the case for the design
For the ultimate complexity of the One Being.

Complexity is found at higher scales,
Not at the original, tiny scale.
Look up and to the future,
Not down and into the past.

Now, something is not just favored over 'nothing'
Because there are zillions of possible ways
To have something
And only one way for there to be 'nothing'.

The case is much stronger.
It is that not anything can become of Nothing.

We observe in our universe
That simplicity begets complexity,
Which begets more, and so forth.

Nature builds complex structures
By processes of self-organization.
Less structure very often
Leads to more structure.

Many simple systems are unstable, that is,
They have limited lifetimes,
For they undergo spontaneous phase transitions
To more complex structures of lower energy,
Such as a drop of water freezing in space.

Barring cosmic rays, this 'eternal' ice
Would last a very long time.

Energy would required
To destroy the structure of ice,
Such as when we use it at room temperature
On our warm Earth,
The heat melting it into less structure: water.

Now what is it that is as simple as it gets? Nothing.
Therefore, we cannot expect it to be very stable.
Note that a total Nothing (capital N)
Could not even do a darn thing,
Since it isn't there, whereas a near 'nothing',
With quotes and a small 'n' represents
The near nothing of the quantum fluctuations/tunneling.

Stephen Hawking says
"...one can show that the negative gravitational energy
Exactly cancels the energy represented by matter."

This shows (and has been shown)
That the universe appeared
From a state of "zero" energy,
This being, of course,
Within the unavoidable
And small quantum uncertainty.

Quantum uncertainty allows
The temporary creation
Of bubbles of energy,
Or pairs of particles
(Such as electron-positron pairs)
Out of near 'nothing"
Provided that they disappear
In a short time.

INFINITE UNIVERSE?

The universe may be but potentially infinite,
As infinities of real things can never complete,
At least by the definition of 'infinity'

The universe would be then only as large
As the outward traveling E/M radiation reaches,
Potentially going on towards 'forever',
As it is a self-renewing wave
That just keeps on truckin'.

CONSCIOUSNESS/ACTION EXPLAINED

Consciousness mediates thoughts versus outcomes
And is distributed all over the body—
From the nerve spindles to the spine to the brain—
A way to actionize without moving.

Quantum consciousness could be behind
The quick collapse of the scenarios of consequences
Into probable/likely actions and/or thoughts.

When a 'thought' finds a solution,
It could just as well be
The nearly instant brute forceness
Of all quantum-type paths being
And evolving in superposition.

When the corpus callosum
Between the brain hemispheres is cut,
Along with some other connecting areas,
To grant relief to those
Having hourly epileptic seizures,
The right and left hand may
Then work in opposition
To each other, neither knowing
What the other is doing.

Sometimes,
One hand will even do something 'bad',
Such as choking one's neck,
But the right hand will come to the rescue,
As in 'my hand is killing me!'.

There are other examples, too, but, in short,
It is seen that the person now has two minds,
Two consciousnesses, and two selves.

Of course, the brain stem still sends common signals
To both hemispheres and to the visual system
And to some others that still inform both sides.

This shows, yet again,
That the brain is the source
Of consciousness, mind and self.

After death?

The brain dies
And takes consciousness
With it to the grave,

For consciousness is
The brain's varying analyses
Taking the stage in turn,
And and is not anything
Independent of the brain.

Consciousness is the arena
For the brain's perception of itself
As coming up with thoughts and feelings
From the memories and associations and all
That that have become us.

A side note is that the brain seeks out all alternatives,
Some becoming conscious and some not,
For it is the prefrontal lobes having to do
With good planning and reasonable behavior
That prune away [bad] alternatives,
As they are seen to be full of mostly inhibitory circuits,
Thus giving us a kind of 'free won't
That lets the [good] will through.

As for what we are,
Neither you nor I are anything special;
We are just brains having thoughts.

BIOLOGICAL CONSCIOUSNESS

As for our theme that consciousness
Is made by the brain,
This is confirmed again since results
Appear 'in it' 200-300 milliseconds
After the brain has finished its analysis,
For this takes some time, as we would imagine.

Biological consciousness
Might arise in trees and such,
But, who knows.

Perhaps a snail has
A limited smudge of consciousness
In which thoughts of just warmth and cold,
Light and darkness surface.

We know that consciousness is of the brain
Because we can introduce
Molecules into brain areas,
This being anesthesia,
And turn off consciousness completely;
Consciousness has no independent existence.

Whatever it is that the brain does
To achieve consciousness
Can be stopped by anesthesia when it dissolves
In the oily regions of the neuron microtubules.

The brain then stays active
But it does not produce any consciousness
Until the anesthesia is taken away.

The same kind of result occurs when you faint;
Consciousness is therefore surely of the brain;
As such, consciousness can be turned off and on
By the xenon or isoflurane gas of anesthetics.

Our environment, inside and out,
Is symbolically represented in the brain,
Our memories and cross-associations
Recognizing and remembering
The meaning of what we 'see', think, feel,
And witness in the unified experience
Of living life as a being.

We all know how wonderful and quick the brain is,
It nearly instantly processing visuals, sounds,
Touches, tastes, and odours
Into higher and deeper systems;

It searches memory very quickly
For what is known,
Such as what the letters
And words of this essay mean,
Forming an abundance of
Further thoughts and actions
Based thereupon, and so forth, and so on,
One hundred billions brain cells
Winking and blinking and connecting,
Making their results known consciously,
At the last, continuing on
In a train of thoughts ever becoming
And arising in almost a kind
Of competition for attention.

Brain cells (neurons) have
A hundred billion connections among them,
Their 'firing' depending on their inputs.

Electricity carries the 'message'
Through the length of the cell
To the gap (synapse),
Where the message turns

To chemical (neurotransmitters)
To take it to another neuron,
Wherein it becomes electrical again,
And so forth.

Your brain neurons have been arranging
Their connections all your life;
It is what you have become,
Molded by your experience and learning.

You are a bio-electrical-chemical being.

HOW THE BRAIN UTILIZES ENERGY AND REFRESHES ITSELF

The brain is amazing and can sort
100 million bits of information in an instant.

Consciousness must play an active role
In the functioning of the brain,
For both evolved together;
They are intertwined in a process.

Consciousness, being a global physical,
Must still yet tie into the separate physical;
So, how is consciousness mediated?

It must be at the synaptic cleft
That what we call 'mind'
Qualitatively meets the brain proper.

Here is where the neuron fires or not;
Here is where a neuron meets other neurons;
This must be where data turns into thought;
Whatever triggers these switches
Produces thought in consciousness.

ZING

Thanks for the vivacity, animation,
Liveliness, vitality, verve, high spirits,
Exuberance, zest, buoyancy, enthusiasm,
Energy, vigor, dynamism, élan, gusto, brio,
Bounce, spirit, fire; movement; oomph, and pizzazz.

ILL-DEFINED 'DEFINITIONS' OF GOD

Here are some "God" definitions refuted
As self-contradicting or are not even definitions:

"God is not definable".

This tells us what God is not,
Rather than what God is;
So, there is no definition
And thus nothing to refute.

The word 'undefinable' even means
That there is no definition.

"God is the universe."

This is a tautology saying only that A=B,
And so we just as well might say
That the universe is the cosmos.

As such, it says that God
Not only does what nature does,
But is one and the same with it,
A very restrictive position.

Another problem here is that this 'God'
Doesn't follow the common usage meaning of God
As a separate being who made the universe
And so this definition
Is somehow trying to slip that in,
But, it doesn't work.

A rose is still a rose by any other name
And so then is a universe.

We already know a fair amount
About the universe
And so at least that part is defined;
It's just that A=B doesn't tell us anything,
As that is just like a synonym.

"God made Himself."

This is saying the contradiction
That something already made
In a design can make itself.

"God is consciousness."

Consciousness is seen to be
The latter portion of a brain process,
Indeed coming 300 milliseconds
After the brain completes its analysis;
Thus it is as far from being some kind of
Initial stand-alone thing in itself
As it could ever be.

"God is perfect."

Again, this is not a full definition,
But only of one trait of one not making mistakes;
However, this is followed by another fact or trait
Such as "God made a mistake
When He caused the Great Flood
And so he invented the rainbow to show that
He would never do it again.".

The statements are seen
To be contradictory;
This happens a lot.

"God is all loving."

Similar to the previous case,
As vengeance is supposedly done on us,
This is not all love.

"God is the First."

Taking God as a system of mind—a being,
Then it is that systems always have parts
And those parts have parts, etc.
And so those lowest parts
Would be among the truly fundamental.

"God is infinite."

This is trying to make God very large
And all encompassing
And thus somehow beyond reproach.

One definition of 'infinite' comes from math
In which one can infinitely divide a finite length,

And another definition is that
Of the counting of numbers being never ending;
There are actual infinities and potential infinities.

This God definition probably means
That God has an infinite amount of energy;
There are then two problems,
The first being that infinities never complete
And so this is saying that God,
A complete being, never completes.

The second is that God uses up
All the energy available,
As His energy is endless and infinite;
However, then it would be
Pretty crowded around here
And there would be
No room left to move around.

"God is one"

God would then be solid,
Like an unbreakable statue
And thus have no pieces and no movement.

"God made the universe."

This is fine, kind of, as part of a definition,
But it begs the larger question of what made God,
And so it settles nothing,
Even adding in an unnecessary step.
It is not an answer at all,
But just a larger question.

"God exists, but is invisible."

This is not known and so it cannot be claimed.

"There was a virgin birth."

Virgins do not give birth.

"God existed forever."

Here we have a case of something,
Actually the ultimate complexity,
Being defined as such
That it was never even defined
In the first place, for there was no first place.

God's earliest memory is then
Forever out of reach to Him
Since there is no first memory;
Anyway, it indicates that a forever
Has already completed
For something real and defined,
But it is the nature of infinity
That it can never complete;
This all goes for all kind of things,
As well, even strings.

"God is all knowing."

Then God faces no surprises,
Which is fine, but is a pretty boring life;
No need then to design a human nature
To see how it turns out.

"God is everywhere."

Then there would be no room for us.

"God detests evil."

Then He wouldn't allow it
Or the Devil to live and tempt.

"God micromanages every quark and electron."

This would interfere with
Their natural movement.
Also, they are always seen to act
According to their properties.

"God works in mysterious ways."

This is a catchall that
Permits God to do anything,
Whether unreasonable, crazy,
Or even against his own Ten Commandments.

"God is outside of time."

Then how was there a beginning of our universe
And a sequence of further creations?

"God is known by faith."

Faith is a belief in the unknown,
Not the known, so there is no 'known'.

"We have free will."

Only if it matches His will.

"God might exist."

'God' is undefined and thus self-contradictory;
What is it that might exist?

LAWS

In summary, religion/God
Was not only not a source of law (but for itself),
But an opposite hazard to progress.

Far from providing us meaningful goals,
Religions prescribe tribal values:
Amity for our tribe, enmity for other tribes
(us vs. them),
Mind-closing faith,
And abject worship of authority.

Thank science,
For it helps us control our own lives;
Divine Right does not.

EVIL

Humans, whatever their take
On the origin of the universe,
Use flawed ideologies
For the worst cases of inflicting suffering;
For, their "concept of good"
Automatically labels any contrary thinking
Or behavior as an "evil" to be wiped out.

SEXUAL ORIENTATION
(From embryo masculinization of Body and Brain)

All embryos begin as female;
If it's to be a male
It needs to be masculinized
In both body and brain,
But, sometimes,
Only one or the other happens.

MaleBrain + MaleBody = hetero male
MaleBody + FemaleBrain = gay male
FemaleBrain + FemaleBody = hetero female
FemaleBody + MaleBrain = lesbian female

MaleBody + FemaleBody = hermaphrodite
(oppositional mixing)

MaleBrain + FemaleBrain = bisexual
(oppositional mixing)

When the Pope discovers
What science already knows,
Such as that all embryos begin as female,
Requiring, for a complete male,
That both the brain AND the body
Must be masculinized (else gay or lesbian),
Then it will be yet another mistake by the Church,
As in the past when it was thought
That evil spirits caused physical ailments
(As well as mental ills, i.e., by the Devil).

AND GOD CREATED WOMAN

God offered Adam a perfect version of woman,
One who would even paint ceilings, cut grass,
Work on cars, take out the garbage, and so forth,
But, this would have cost Adam an arm and a leg.

So Adam said,
"What can I get for just a rib?"

WHAT WAS IT THAT WAS AROUND FOREVER?

Could a Mind, such as we see in humans,
Have been around forever?
No, of course not, for it is an already defined
And very complex composite system.

Ok, then, how about the fully formed
Cells and molecules of which the brain is made?
No, still ordered and complex.
And, yes, I'm stating the obvious.

What about an atom
Or even its smaller constituents of?
No, not even, for these parts,
As small as they are,
Have some definition of properties—
These parts have a specific order:
A size, a hazy spin, energy, etc.,
And a timely 'when', a specific 'where'—
And even how many of them there are.

If they had been 'forever',
Then there would have been
No first place for them to have been defined;
Therefore, the realm of the causeless
Must be one of no definition at all—none whatsoever—
An utter and complete shambles of chaos,
Called merely 'potential' or 'possibility' here;
That is, a lawless, formless, 'anything goes' realm
That precedes real form and its movement and laws.

Such, indeed, is what we see
Of the virtual particle pairs
Arbitrarily arising from the quantum domain,
As in geiger counters randomly beeping,
All being the indeterminate
And random happenings of disorder.

So, while order always needs cause beneath it,
The disorder of chaos does not.

All springs from the uncaused ground-state,
Vibrating forth these particulate songs of reality,
From which all complex composites then form,
Our brains being the most complex, so far.

THE CAUSE, JUST BECAUSE

Some theists say: *By all appearances,*
Life's chemistry looks like
The product of a Creator.
A nontheist might respond:
Life is but a tiny spec
Compared to that of the Creator's,
So, by all appearances, then,
To be consistent in demanding cause,
God's much higher life, all the more,
Looks like the product of a another,
Much Higher Creator.

But, *no,* it might then be said by the theist,
Some Lives require no cause.
OK, fine, thus the infinitely simpler life
Of the universe needs no cause,
Then, all the more.

So, we see that these kinds
Of mere 'pronouncements'
Attempt to halt all further inquiry
Such as the saying of 'God did it',
And indeed it may do so for theists;
Whereas, in science inquiry is ever welcome
In the search for answers
Beyond 'case closed' and 'just because'.

KNOWING REALITY

The use of accumulated knowledge
Is used to make objects by engineering
That actually function as intended.

Since we can make objects
With predetermined behavior consistently
Means that we have correctly
Understood external reality.

SHAKESPEARE GOT THE GENERAL IDEA:

Such tricks hath strong imagination,
That if it would but apprehend some joy,
It comprehends some bringer of that joy;
Or in the night, imagining some fear,
How easy is a bush supposed a bear!

QUARRELING

The lively couple meandered on,
Even through the ToeQuest threads,
Fancying that they were not alive,
But then smiling because they were.

The hours were fresh and mild,
Like cleansing showers,
And so the partners could retrieve all
Of the wingéd hours that time
Had attempted to devour.

As they walked, the peace of the forum
Was shattered by the sound of someone
Bickering and quarreling
In an all too common way.

He and she approached the noise.
The squabbler just stared at them at first,
But then started arguing
About anything and everything.

"Save your breath," she urged.
"Don't expend it on fighting and arguing.
Fighting saps your energy
And forever undoes love's promise;
Your breath is dear and your breath is precious;
Enjoy all that life can give, ere comes death.
Yelling drives people away;
Soft and gentle voices, whispering even,
Brings them closer.
Tell him more, partner."

Her partner continued,
"There are large worlds of life to live in.
But, here you are,
Trapped in a little tiny cell of arguments,
Resentments, and animosity,
Wasting all your breath therein;
Stand back and realize life's total space—
And note that quarreling occupies
But a small place in that
Which can be accomplished
By the human race."

The hopeless criticizer kept on ranting,
Getting mad at every single thing;
If s/he ever had anything important to say,
It was now lost in the quagmire.

She sighed, "Well, then, if you're not busy living,
Then I guess you're busy dying.
All the world's riches cannot extend the power
Which drains the cup and withers the flower,
So, what would be the price
Of even your wasted breath,
Purchased from the hand of death
At the final hour?
Loving is what this life is all about.
To have it is to live all out.
Then why, oh why, do you not seek it out?"

The oblivious quarreler kept on arguing,
Bickering, and criticizing.

THE GIFT OF BEING WAS NOT A

Present from a Giver,
Something that was handed out,
Donated, an offering, bestowal,
Bonus, award, or an endowment.

The gratuity of our largesse,
The so-called perk and benefaction
Of our being was a benefit
Of the way the universe was/is—
The luck of our path through evolution.

It is unavoidable, then, to 'knock' religion,
As truth comes out, and then,
Even more directly when asked,
As the truth is explained and shown,
The falsehoods necessarily receding...
Reformulating into a smaller
Space for God to act...

Until...
There is no space left
But that of a dice roller...
And, even then it can be shown that
What is without cause is all that it must imply:
That the Great Designer Himself
Could never have been DESIGNED,
This thought experiment confirmed
By the indeterminate chaos
Of the quantum realm.

HAS THE MULTIVERSE
FINALLY BECOME NECESSARY?

Long ago, protons had formed
A massive star, via gravity,
And for quite a while
It had fused hydrogen into helium,
Living a long and healthy life,
But its death would be even more spectacular.

In its death throes,
This massive star goes out with a bang—
First collapsing, then triggering
A supernova explosion bright enough
To drown out the light of an entire galaxy.

A shock wave of precious stellar debris
Hurtles outward into space
At tens of millions of miles per hour,
Containing the heavier elements
That will make up planets,
Create more stars, and even create life.

Well, here we are,
Being made of the atomic elements—
Stardust, or chemicals, if you will,
And to this dust we'll one day return,
Like it or not.

Suppose some of those protons
Had been just 0.2 percent larger;
They'd have been unstable
And would have decayed.
No atoms would have existed, ever.

Suppose gravity had been slightly more powerful;
Then the stars would have been smaller,
Being more compressed,
And they would have sputtered out
Long before they had a chance to evolve.
No life.

Suppose the electron
Had been twice as large;
No life.

Suppose that,
When two hydrogen atoms
Were fused into helium in a star,
0.006 or 0.008 of the mass

Was turned into energy (E=MCC)
Instead of 0.007;
The universe would then have been filled
Only with hydrogen, if it had been 0.006,
Or their would have been
A universe with no hydrogen,
If it had been 0.008.

Suppose the early universe
Had not a delicate balance,
But had headed into runaway expansion,
Much too quick for galaxies to form,
Or had headed the other direction into implosion.
No life.

Suppose matter had been more evenly distributed;
It would not have formed galaxies.
If matter had been clumpier,
It would have all condensed into black holes.
No life.

What if the strong force
Were slightly more powerful?
Then all protons in the early universe
Would have paired off
And there would have been
No hydrogen to fuel long-lived stars.
No life.

It is because we are here
That we can now look back to see
That the universe was made for us.

We had to be in a universe
That had reached a certain age,
A certain stage of evolution
After stars had formed,
So that life could arise.

It is that in our universe
The odds for life were just right,
As they had to be,
Or we would not have been here
To marvel about it.

Inflation might be an ongoing process
Throughout the universe,
Where even now some different regions
Of the cosmos are budding off,
Undergoing inflation,

And evolving into essentially separate universes,
Then doing the same again, eternally.

The expansion of the universe
Is even accelerating;
Something is pushing everything apart;
Yet, it's not too fast and not too slow.
Too much dark energy and then gravity
Would have been overwhelmed.

This particular fine-tuning
Seems to be extreme luck.

Our universe may have been one
Of perhaps infinite universes,
Each with their own laws of physics...

Now, while string theory seems to be useless
In the way that it cannot prove anything,
It does have 10**500 solutions
Of unique ways to produce a universe,
Meaning that a multiverse
Might spawn real universes
In many of those possible ways.

Our universe could be
Just one of a multitude,
One that just happened
To have the right kind of physics
For carbon-based life.

It seems that the "All"
Has always gotten larger
And that this is the next step
For our realizations
And so our universe may not be
All that important in the scheme of all things,
It being but a tiny spec in the overall multiverse.

Once we thought that the Earth
Was the center of all, then the sun,
Then the galaxy, then the universe,
But now... the multiverse!

We are so small,
So tiny—a spec on a spec
On a mote of dust.

Did the multiverse come into being
So that our one flower could bloom

And blossom for just a moment?
I doubt it.

Are all the universes real
Or does this all happen in a kind
Of "Potential/Possibility" form of superposition?

Also, how is it that the many universes
Can have different physical laws?
It must be a dice thrown from chaos.

Did 'something' have to be?
Yes.

Why?
Otherwise,
Nonexistence would have
[Been all around],
But, it could not.

Thus, some things forming
Is the natural state of affairs.

A total Nothing could not produce anything.

The 'something' is the near 'nothing'
Of the quantum fluctuations/tunneling.

Could all real things have been around forever?
No.

Why?
Forever could not have already happened,
For infinities never compete,
Plus, all would have been worn out—ha-ha.

Most importantly, how could something real 'be'
Without it ever being made?
What would have determined its amount,
Where it was, why it was, how it was,
Its nature, and so forth.

What, then?

Only the potential and possibility
Of real stuff could have been around forever,
It being the necessary
And complete chaos of the causeless,
Thus needing no more regress
Toward the even smaller to be made of.

It could even evolve all possible universes
By the mindless brute force of possibility
Perhaps the only way
To 'get' some of them to work.

Why?
A Natural law is proposed—
The natural extension
Of "physical" dimension
Beyond the actual 3-D;
It's kind of like
The everywhere of superposition
In the quantum realm,
Only this would be all possible
Universes in superposition.

Some think that quantum events split reality,
One in which, say, they die,
And one in which they live,
As an extreme example.
Who knows.

What I like about the concept
Of other universes 'trying' to be,
Is that perhaps some made it, like ours,
And some fell flat,
Which could explain how some
Of the 6 or 7 precise universal constants
Came to remarkably be just right.

Now, on the other hand,
Victor Stenger doesn't think that the universe
Is so finely-tuned as we think it is,
And if some of it is for carbon-based life,
Then it is proposed that another kind of life
Could have arisen if this were not so.

Thus, we were also fine-tuned to the universe
Rather than completely the other way around.

PREVENTION

As premarital sexuality is not a natural state,
According the Church, it doesn't approve condoms.
Yet, the natural state persists, as it ever did,
Incurring sexual diseases, sometimes.

There were other diseases before HPV/HIV.

THE MAGISTERIA MUST OVERLAP

I can show that the magisteria already overlap.

1.

Suppose that there is an imagined supernatural Notion
That is supposed to have effects upon the natural,
Even luckily being that the Notion's effects
Are to be found anywhere and everywhere.

Science, as it goes about
Its normal fact-finding business,
Probes and examines anywhere and everywhere,
Ever only finding the natural, that is, no violations,
No super-beyond-anything happening at all.

The conditions thus look exactly as they would
If there were no supernatural Notion.
There is then inadvertent overlap;
It's unavoidable

OK, that's an easy one, even a double-demise,
For the Notion couldn't be proved in the first place
And so, thus, even further,
Is of no concern of any consequences from disbelief.

People make more informed decisions lately,
Good probability being enough,
As there cannot be complete
And perfect information.

2.

Now, nonexistence of a Notion, effects or not,
Can also be shown
If the Notion is self-contradictory;
No square circles;
No Design without DESIGN as a first cause, etc.

I leave this to the readers.

3.

So, onto the tougher case that occurs
If the nonscientific magisterium retreats [giving up]
To a new position that there are no effects,
There being no intervention at all
But for the Notion having
Just created the universe
And not intervening any further.

Now, note that science is still but doing its own thing,
Investigating more and more fundamental realms,
Such as even proving that the quantum level
Is a random and indeterminate chaos,
That there can be no local
Hidden variables within it;
Thus, finding even that a near 'nothing'—
The quantum fluctuation or tunneling—
Is the causeless bottom 'something'
That is as simple as it gets,
This 'something', by the way,
Being the natural state of affairs,
Rather than a total Nothing that could not be,
Confirming the thought experiment
That a total Nothing couldn't do a darn thing,
It not even being able to be 'there'
To make anything anyway,
Plus, that there cannot be
Never-ending causes beneath causes,
And that therefore the causeless bottom
Must be of maximum non-specifics
Rather than any order,
Much less a perfect order.

Thus, and, too,
The causeless bottom needed no creation.
The Notion is not only cut off at the source
But is not even required
Since the normal state is 'something'.

The great philosophical question of
"Why is there something rather than nothing"
Is squashed,
For not anything can become of Nothing.

Science even then finds, as a bonus,
That the universe appeared
From a state of zero energy,
This being, of course, within the unavoidable
And tiny quantum uncertainty,
Plus, that the negative energy of gravity
Matches the positive energy of matter,
Equaling a mass density of 'zero',
And, further, that every time
We try to measure what an atom does,
We get a different answer,
This then again being the answer
That that realm is causeless.

Furthermore, that realm
Is of discrete operations—
The quantum leaps
Even wiping away the notion
Of any universal continuity.

Another bonus found is
That the 'laws' of point-of-view invariance
Automatically appear [are not handed down]
In any model that does not single out
A special moment in time, position in space,
Or direction in space,
Such as back at the Planck time
Of the big bang,
The universe having then
No distinguishable place,
Direction, or time, meaning that
It had no structure
And, thus,
That the conservation laws apply.

Further, it can be shown
That human and societal behaviors, morals,
Laws and values look just as
They can be expected to look
If there are only the natural goings on.

Science, in its quest for truth,
Has inadvertently stepped on the turf
Of the nonscientific magisterium.
There is much overlap.

THE MYSTERIOUS SCHOOL

I looked for the ancient mystery school;
However, its whereabouts remain a great mystery!
I looked deep into the unified field
For the hidden knowledge
Of our genetic encoding
Of 7-dimensional consciousness,
But it was really too well hidden!

I looked for my tennis shoes
And there they were right under the bed!

GALILEO: REVELATIONS

If Galileo had noted that Mars
Increased and diminished in size during its orbit,
Then this could have been a good clue,
Although perhaps not a proof,
That the planets orbited the sun.

The Church wished him to just say
That he only had a hypothesis, not a truth.
(Parallax measurements were not yet known.)

The real concern of the Earth
Not being the center of all
Was that perhaps Hell was not to be found
Within the bowels of
The no longer so important Earth,
As well as the concentric crystalline spheres
Surrounding it not being so,
Although this notion was really
Only proposed by Dante.

And, too, for some reason,
The notion of the universe being infinite,
As perhaps then
There would have been no room for God.

It was also a time of challenge
From the Protestant reformation
And thus the necessary
Catholic counterreformation;
Galileo was in the right mind
At the wrong time.

One could say that Galileo
Should probably have known about
The sensitivities of the Church,
And probably did,
But perhaps felt safe
In his association with Pope Urban;
However, it was also that
The Pope had to perform his job.

The last straw that broke this friendship
Was when Galileo portrayed the God argument
Through the mouth of 'Simplicio' (the simpleton),
Even presenting it much
As the Pope would have to present it
And actually did present it to Galileo
Once upon a time.

It was just that Galileo had come upon
A great secret of the universe
And so, like anyone,
Could hardly contain himself.

He performed quite a balancing act,
Even stating, perhaps as a deflection,
That his argument was with Ptolomy,
Not the Church.

DREAMS AND REALITY

Night-dreams are not the same category
As when one is awake, but, obviously,
The same simulation model
Of reality is employed.

In a night dream there are
No sound waves coming in,
But there can be sound,
No E/M waves coming in,
But there can be light and vision, etc.

When awake, there are both the inputs
And the representations.

Dreams

FREEDOM

The fact of there being no purpose,
Turned inside out,
Is to know that we are free
Of the puppet strings of commands,
Leaving us free to enjoy life
And make our own meaning out of it,
At least one confined to this particular path,
Within the limits of our form therein, of course.

CONSCIOUSNESS MOO-GOO

Thinking, feeling,
Awareness, and consciousness
Are processes that depend
On the constitution of the brain;
They are not stand-alone,
Absolute free floating things
Roaming freely around the universe.

Take the process of conscious awareness;
It depends on a brain:
Alter the brain and consciousness is altered, too.
Add some molecules of anesthesia
And consciousness is even prevented.

When you are very low on energy
Or have been awake a long time,
Consciousness fades and you fall asleep;
There is no conscious awareness without a brain.

Some think that since
The process of consciousness
Is somewhat mysterious
They can just pronounce it
As being some First kind of thing,
Such as a spirit that pervades
And creates all else.

Well, consciousness is not a First,
But is, in fact, rather,
Near the tail end of a process;
Nor does it create,
But is in fact itself created in the brain.

One cannot just take a process
And turn it into a magic word
And then simply pronounce
That it goes on all by itself
And does things without
Its underpinnings around.

THE DREAM OF REALITY...

I would suppose that the only way
For the Dream believers to have
The Dream be true is that
Nothing is really 'out there'
Except for a Dream-Projector mechanism
Which presents an awake-type dream
In which every interaction
That we only think we observe for real,
Down to the trillions of them
In every little spec of the universe,
Is simulated in its absolute and total function
And is thus played to us 100% exactly
As if it were really real.

So, then, D=R, is a tautology,
And also a 'difference that
Then makes no difference.

Fine, (not really) but remember
That the optical system and the other senses
Can then take nothing in
Since nothing is out there;
So, all that machinery is just fake,
Really doing nothing at all, just deceiving us.

Who or what is perpetuating this fraud?

No one can know, since it is all... um,
Invisible, but then how it is fact?

BEHAVIOR

"Curses Christmas!" said old Scrooge to his yule log;
"Why my hard earnings give to some lowly dog?"

"'Twere all these logs that you stepped upon,"
Said Christmas Past, "enriching from these anon,
Building here your own little private town."

"Lo", said Christmas Future, "There's no one around!"

UNCENTERED

The Earth was found not to be the center of all,
Nor the solar system or even the galaxy, and
Now it is even that it could be that our universe
Is not the center of the multiverse.

When people climbed Mt. Olympus
And saw no Gods there,
That mountain of myth crumbled, too.

That the connect-the-dots Gods of Astrology
Were not to be was another crack
In the foundation that violated the building code.

Then their are more indications
That the complex derives from the ever simpler,
The ultimate simplicity now
Being seen as the causeless,
For there are not infinite causes beneath causes
Of that 'something' that always was;
So, when there is no creation,
There can be no Creator.

None of the above gave pause
To the religious who attacked the truth,
Such as with Galileo,
But, the Church could not burn the truth away,
For the proof remained.

'God' is just a theory;
However, the Church's deception
Is, as in fact, to speak of it as the truth.

HERE I AM

Before I even was, I came here to be,
The energy of eternity being my maternity.
I found myself lost, really, here without asking...
But, eventually, at the lost-and-found,
My karma overran the dogma into the ground.

ALL IS ILLUSION?

The problem,
With the saying that nothing
Is really 'out there'
And that All is an illusion,
It being a nonsensical babbling
And jabbering of the Dream talking to itself
With no real significance is
That one then sometimes uses
What's 'out there' to support it,
Such as old Indian writings.

By the claimings of illusion,
There is no real book out there
With any significant information
Whose pages one is turning
To gain a deep insight;
It is only the Dream
Producing a fake picture
Of a fake book's pages turning
To make it seem like it
Is out there and being read,
But, all is gibberish.

The same for drinking wine, swallowing food,
Digesting it, and gaining nourishment,
For it doesn't really happen;
Only a picture of it happening is presented,
And so forth, and so on
Through all real things discussed,
From referring to evolution or to the earth
Or that the earth is 3000 years old, etc.

If all is illusion, then one has
No firm place to stand
To even say that it is all an illusion.

THE 'HOLD UP'
(Stolen from Fredrick)

A belief is that construct that states
We consider something as true;
But considering and knowing
Are two different words.

One implies holding something up as true,
While the other stands on the ground as being true.

ALL

'Shortly' after the birth from light
Particles congealed,
But they left an imprint on the light.

The wrinkles of early space blew up
To form clusters, galaxies,
Stars, planets, and us.

We're used to it.

When the moon throws
Its cold shadow on us
From an eclipse,
It shatters the place
That we've gotten so used to.

By a rare coincidence
The tiny moon snuffs out our star
When the sun 'sets—at high noon.

The history of science
Is a deep lesson in humility—
We have no privileged spot;
We are of the ordinary.

Even the universe is nothing special,
Probably just one of many.

Our ancestry can be traced all the way
Back to bacteria, of all things.
And we still depend on them.

We still share 50%
Of our genes with fungi.

Our emotions boil down
To dances of molecules.

Profound thoughts
Are neurotransmitters blinking.

We don't amount to a hill of beans.

Everything flows from atoms;
Chemistry drives all.

Nature doesn't give a hoot about
What humans think is common sense.

Man is hardly the measure of all things.

One may ignore the truth
To further their aims,
But, Pandora's box of info has opened.

Let us then have some shred of humility
Rather than forming schemes
To make us seem so special and deserving—
We that have flowered but for a moment
As the famous Graybeard would say.

DEFINITELY

If it is that definite formed things are made
And thus are not eternal,
Then there is still the timeless
Basis of them that was/is,
And even the preceding words not withstanding
It is that the causeless basis
Of potential and possibility
Was always there,
And has never gone away, for it can't,
Being that it is the normal
And natural state of affairs,
Rather that Nothing,
Which can't even have existence to be.

So, it is still here somewhere,
Maybe even everywhere,
As perhaps in the quantum foam
Of the vacuum energy,
But was not this, too,
Created along with time and space?

If so, then perhaps the foam
Is still some kind of shadow state
Of the original potential,
It emitting particles everywhere,
And shown to have no hidden variables?

Are there some rare circumstances,
Say once in a zillion years,
Those that are just right to produce
An unusually massive outpouring of material?

WHAT WE KNOW

Well, let us look at what
We do know about the original
And necessarily causeless fundamental basis of all,
Whether it be energy, substance, or potential,
Since those are 'just' the details
Of its form and implementation
That get worked out more and more.

What can we get out of that
Which we know for sure,
Ideas that mankind at large can use?

What is the message,
Rather than the mechanics
Of the messenger?

It, whether one
Or several separate fundamentals,
Rearranges and builds,
From the incredibly tiny and simple,
Through recombination,
To the larger and more complex.

So, the fundamental itself
Has no attributes of the complex,
Such as any kind of composite system all foreseeing,
Which, by definition, would not be fundamental

Yes, the nature of the attractions and replusions
Is always of the basic material.

Thus, greater things lie in the future,
A good direction in which to look,
Rather than backward.

Look beyond, rather than beneath.

So, one may invent any purpose
For themselves that they wish,
Since the atoms, electrons, quark jets
And other small building blocks
That we do actually see
Are not marked as good or evil,
Holy or unholy.

The fundamental
Was never created,
Of course,

It being fundamental;
So, there is no Creator,
For the fundamental basis
Was never created,
Which, of course, if it was,
Would imply something before it,
Making it not the fundamental stuff.

(Intermediate, middleman aliens
Creating anything
Are not fundamental,
But are complex composites.)

So, superstition can be dispensed with;
Thus, there need be no angst
About the fear of being condemned to burning;

In fact, there is a joy of freedom
From having no strings being attached
(But for those of the limitations of our form).
It can be rather exhilarating.

Thus, the results of the fundamental,
Us and our universe,
Are neither right nor wrong
But only that it functions.

We don't know if it came
In other variants that worked differently
Or didn't work at all, but, no matter.

So, there are no black and white answers in life,
As our lives depend on the nature of the stuff
And what it has become.

What's relatively 'good'
For one part of nature
May be 'bad' for another part.

For example, some bacteria can cause disease,
But we can't live without bacteria.

Or, without death,
For evolution would not have progressed;
So, death is necessary and natural.

It still could happen that
We eventually conquer death somehow.

Atoms have great energy, a boon,
Yet, atomic bombs can be made.

Aside from the information of the above,
Existence trumps and precedes essence,
In importance, by far.

This is because we MUST
Deal with our existence,
And could even do so
Without any knowledge of the essence,
Although we do have some
From what we know for sure.

In general, one is free
To make one's own meaning in life.

Whether the fundamental could have
A definite design of form is debatable,
But we'll continue to go beyond
Those details unknown,
Concentrating only on the known.

Since it always was,
And it still is, and ever will be.
Life, in some form,
Will ever arise and continue.

The fundamental 'something'
Is the natural state of affairs
Because a total Nothing,
Meaning a lack of anything,
Did not and could not occur,
Obviously, since something is here.

Thus, there need be no angst
Over why does anything exist at all
Or why is there something
Rather than nothing,
For those questions are stated backwards,
As if Nothing could be
Or do anything at all.

ZERO

MY SEARCH FOR THE TOE WAS AIDED
BY THE SEARCH FOR GOD (FINDING NONE)
OR VICE-VERSA.

I first ran into Emmy Noether when trying to see
If the laws, morals, fine-tuning,
Cause, tine-zero imprint, etc.
Of the universe were natural,
Rather than extra-, or super-.

Since the Theity is said to be everywhere
And to intervene everywhere,
The lack of evidence that is supposed to be there
Would truly become evidence of absence.

This research went beyond the obvious case
That the supernatural beliefs
Had no ground to begin with,
Just hanging in the air as unfounded.

While that is ever still the case,
It didn't stop people from preaching theory as fact
And so my research continued;
So, now we have such religious proclamations
Being two steps removed from reality,
One step to ignore what is,
And the other to pronounce what isn't.

In short, I found everything to be natural.
Morals, values, and civil laws came from humanity.
Physical laws were naturally explained by Emmy.

No time-zero Imprint was found,
But, rather, just the opposite: chaos.

Biblical revelations,
Those not written after the fact,
Did not come to pass.

Atoms gave a different answer
Every time we measured them,
A behavior apparently causeless.

Alain Aspect showed that
There were no hidden variables
Within the quantum realm,
Unless superluminal.

...

The fundamental, too,
Having nothing prior to it, must be causeless,
Which astoundingly revealed
That there was no Creator of any kind,
Not even a Deity,
Since nothing was ever created in the first place!

This cuts off all other
Possible 'God' arguments at the source.

The SuperToe is simply that all was causeless,
As causes beneath causes cannot go on forever,
And as Nothing cannot produce anything.

The regular Toe after that
Is how things function beyond that,
Given the original substance, energy,
Or quantum possibility,
Whichever mechanism it is.

Could it be that in the original underlying basis
Beneath strings or whatever
Is that 'anything goes',
There being no firm establishment
Such as the particular structure
Of our 'secondary' universe
But that it is one of the possibilities?

Everything possible would be superimposed there.
Do we know of such a state?
Yes, such we see for a quantum entity like an electron
Being everywhere but nowhere in particular.

So, here we may be, either in a firm particular universe
Or even in all of them,
The probable state dominating most of the time.

Would anyone venture to say
That a very likely extension
Of our established 4 dimensions
Would be that of all possible universes?

Does this make our universe good
Only in that it works?

Or, further, that it is morally
More right than wrong?
Or is it not a factor
Since the established material
Knows nothing of morals?

The paradox is still that
Some apparently definite and fundamental stuff
Has a certain finite number or amount of base,
And, yet if it's fundamental,
Being always there and necessarily causeless,
Then there wasn't anything
To define it in the first place
Since there wasn't any 'first place';
So, something is wrong,
Plus, we hate unresolved paradoxes.

Also, if something really big
Did happen 13.75 billions years ago,
Then we should at least look into if it
To see if it had an origin
Or if it was just the same old original stuff
Returning somehow in a cyclical manner.

If the next WMAP finds no gravitational waves,
Then at least inflation is out of the picture
And the way is perhaps paved
For a cyclical universe.

Might we then,
Though in the first. 'origin' case,
Have to go beyond energy,
Just as we've tried to go beyond matter,
Seeing as there is an equivalence?

So here, then, we stand
At the edge of an abyss,
Unless we know a way further.

What about the arrow of time?
Nothing says it has to have a direction.
It's just unlikely, but possible,
That air could go back into a tire.

Is the original underlying basis
Something like a fluctuation
Or even a universe happening forward
When another goes backwards?

This, although some plus and minus energy
May cancel out to help with the paradox,
Is not something from nothing
Since the event's capability
Had to be there beforehand.

+/-

HIGHER AWARENESS

It's not that we're stoic,
Totally ignoring the world's problems,
Or totally devoid of any feeling
Or being effected by 'life',
For there is always what's best for all to be done.

It's just that we can carry on with progress
Without becoming really greatly disturbed,
And, sure, if one is fighting for their life,
Then one can surely use all the adrenaline
That one can summon up.

Our influence increases with
Those things around us
That we can some large effect upon,
Such as ourselves and our reactions,
For example, which, in turn, will flow beyond us.

I need to use a negative, here, now,
To demonstrate the usefulness of the positive
And even the neutral position:

Say that by accident
We spill a whole gallon of milk
All over the place,
Perhaps even shattering the bottle
To make it a worser case.

These things happen in life,
And are bound to,
The only real 'surprise' ia of
Being when it comes to pass.

Well, at any rate, the milk has to be cleaned up,
For we must deal with our 'existence';
However, it is completely optional
That we go nuts, scream,
Jump up and down, swear,
And otherwise increase our cortisol level;
In fact, this may even impair us
For the job of cleaning up the milk.

Now this is not to say that there aren't minor
And involuntary psychological reactions
That peep out, waiting to mushroom into a bomb;
However, some do let it escalate.

Of course, these kinds of people
Never have much peace in life, for it gets worse.

Already haven gotten
To the unnecessary stage
Of irritation and anger,
There could then be a run-in
With the next person who happens by,
Or even from some otherwise normal type of noise.

The brain is not so smart, sometimes,
As it is wont to pair the bad feeling
With the most recent event,
The noise, say, and rail at it,
Increasing anxiety levels even more.

Then, for some, it still may not end,
For everything in life then becomes
An immediate annoyance, and so forth...
The world is against me... everything stinks...

Or, even worse:

— The (D)anger Zone —

Once we drop into the anger zone, the
Analytical mind cuts out, giving way
To the primitive reactive mind, a
Moronic state in which even beige seems black.

— Thus, All Become Equal —

The simple reactive mind 'thinks' that, say,
A perceived bad tone equals insult equals
Hate equals great anger equals lash out
Equals big fight equals kill equals death.

So, believe it or not,
I've said something positive here,
Or at least neutral,
Until one gains enough practice.

WHAT GOD WANTS?

I don't think God asks us to be holy,
He asks us to become whole, decent, loving, fair,
Just, and to live with mindfulness towards one another.

Hmmmm... I do this naturally,
But it is an option of Nature as it is,
Not a forced direction, so, again,
That is but said to be His intent.

It seems that not all people have this full capability,
That all are not created or raised equally,
That life is not fair in this regard.

Well, life isn't fair and that's the way it is.
I'm not sure where anyone
Gets the notion that life is fair.
All the should-be's for life being another way
Just don't count a bit in the face of the fact
That life is just what it is in the fairness arena.

So, to continue in our 'as if' analysis,
Although first noting a lesson
That whatever Austin or anyone makes up
About the nature of God, speaking for Him,
Or even that He is, is, in the end, still made up.

We are not dwelling on this here,
For we are doing the best of 'what if',
But it is a lesson that should not be lost on anyone.

So, I would certainly have to agree
That the Scientific Deity type God
Is the only one that is possible,
Being that He does exactly what nature does.

The supposed 'Theity', as known,
Is but a lot of window dressing
That humans clothed Him in.

And, too, no straightforward
Beyond or extra or super kind
Of evidence exists for the Theity
Who is supposed to be everywhere,
People's inner sensations
Of their minds not withstanding.

Ah, but the Deity set material in motion
And also constructed its nature to carry on
With His Master Plan,
Which, by the way, was not fixed like a DVD playing
But purposely allowed for variation on the theme.

So, if we may make up more assumptions,
Such as the Deity has being,
Then the Deity just sits around,
Even now that his plan is in motion,
Doing we know not what,
But, there are some questions
That almost give Him pause.

He almost wonders how He got to be there
In His position of ruling All
But then of course realizes
That He never 'got there'
But was always there.

He is not worried that
He has no earliest memory,
For there is a magic feature that allows Him
To search infinitely back
Into all the happenings previous.

He is almost stumped on why He is there,
Like 'Why Me?'
But since His 'something'
Is the natural state of affairs,
Rather than the unproductive Nothing,
He comes to terms with that near 'paradox'.

As said, He is so infinitely intelligent
That there is of course no vengeance
Or any bad emotion happening in Him.

Again, humans made Him be more that He is,
Molding Him in their own limited image.

The only real question that might give Him full pause
Is how He Himself is already there fully formed
In His extreme state of complexity.

Nevertheless, He builds the plan for a universe
That works exactly as designed,
Doing it in less than an instant, not six days,
For there is no limit on His power,
Nor does He get tired

And have to rest for a day,
For His energy is infinite

So, this Biochemist Scientist
Is the Ground-of-Determination (G-O-D),
Although stripped of all
The human holy-moly painting
Of his character and His aims,
Even including the assumption
Of His being a complicated
Complex composite being
With a magical built-in system
Of a totally functioning mind.

When you really get down to it,
After the necessary minimization
To being Nature itself,
He is the TOE.

CONSEQUENCES

Leave your bike out in the rain and it will rust—
A natural effect and consequence.

If you live on a corner and let hedges grow wild,
Car drivers may not see across very well and crash,
You then getting fined—a logical consequence.

If someone disrespects property
And throws a regular chair into the pool,
They, although some might think so,
Wouldn't see all the possible problems,
Like maybe its get cold and no one can swim
To get the chair back, or whatever.

In general, if one, say, has a general idea
That respect is a good thing,
Then one is not burdened to think
Of all the possible consequences
That might ensue from disrespect,
No one being able to see them all anyway.

NOTES ON NOTHING

If something could be made from Nothing,
Anything could and would
Spring out anywhere, anytime.
Nothing

MASS AND ENERGY

Energy/mass transformations
Are the physical substance of the universe.

Since waves make particles
And particles behave like waves
Shows that E=MCC...

Lots of energy for small amounts of mass.
Mostly empty space is the way nature works.

The brain puts a much better face
On these frequencies "out there"
And so we can get around
The universe all the better.

An example is that some wave frequencies
Are seen as color in the mind,
But the rest of the electromagnetic frequencies
Are not seen except by our instruments.

SIMPLICITY

The 'probability structure' is simple,
It being a blind type of brute force traversing
Of all possible states/paths,
Eventually, or maybe even all at once,
According to some probability,
Much akin to the quantum happenings.

It seems that religion and science
Must ever converge at the causeless,
One expecting perfect order,
The other finding complete chaos,
Or, at least, no specific definition.

It's difficult to fathom
How any certain definition
Could have been imparted
To the causeless ground-state.

At any rate,
Existence must always
Precede essence in importance,
By far,
For what can a claim upon
The invisible really do for us?

ON DESIRING THE INTELLIGENCE, CREATOR, DREAMER, ETC.

I notice that an (in)vested interest
In the unseen Creator,
Also known as the Dreamer,
Causes much of reality
To be turned inside out
To reach and protect
The 3 or 4 such religious paths
Of claims upon that
Which is unknown and unseen,
As well as leading to some heated words
That kind of say
"How could you possibly have a claim
That differs from my own?"

And this is not to mention the other claims
Of scientific non-religious facts
Based on what is known and seen.

The first position or claim—
I'll pick one at random—
Says that the brain does little or nothing
To direct the person,
Which thus allows for
The soul to do the same.

It's no use having two things giving direction;
So, the brain, expensive as it is,
Has to go into the organ donor bin.

The soul, then, thus established, leads to God,
Now the "actual" of the vested interest.

Evolution, too, must not only be also discarded,
But, worse, it must be said to have never happened,
Not just because it would have led
To the brain evolving to actually do something,
But also because man would not then
Be the measure of all things.

They do, though, posit
That all excepting the soul/mind
Is physically real.

The second claim position also admits
To the realness of the physical self,
But says that the ego must be banned,
Either because nature got it wrong,

Or, even got it right
By intending it to be a very prominent,
Nearly unavoidable deception, so,
This, too, then makes way for the soul and God
To manage the person
Who then becomes closer to God.

The first position then comes back
To declare that some part of the brain
The ego—indeed must do something worthwhile
And so it should not be
Banned at all in any way,
Being absolutely necessary,
And perhaps directed towards good.

Each declare the other wrong,
And, indeed, both can't be right,
Nor might either one be.

At least the first two claims
Have in common
That there is an individual
To choose to do something
That can have a real effect.

The third position claims
That there is nothing physical at all,
Nothing existing but the illusion of it,
This claim really turning the universe
Inside out to have its way.

The Dream goes on as it does,
With behavior good or bad
Being just a part of the babble,
Neither wrong nor right,
But just being what's
Playing in the theater,
No individual really existing
In its own right to have effect.

It all three cases the desired conclusion of the end
Is employed to direct the 'authority' of the means
Of getting to that exact end;
Plus, all three are really a great stretch.

That they vary reveals the great wish
To make the (in)vested interest be so;
That is, just about anything and everything
Is thrown away in order to get there.

Emotions are often hurled at those differing,
As if that adds any value to the claims
Of where peace comes from.

A kind of 4th claim,
Is to turn human consciousness,
A tail-end process,
Into a first and primary 'thing' somehow,
It now becoming an actual free floating entity,
One that perhaps builds the physical
Which then builds the brain
Which...has consciousness
As a witness to what's going on.
Circular?

And, so, from the (in)vested interest
In some invisible imaginaries,
We have several conflicting views
Of reality just right here,
Although there are many more out there;
But, only one or zero can be right, not all.

It is that the (in)vested interest
In the invisible imaginary
Steers one's notions of reality,
Here even in four different ways,
But it is really that the claims upon
The invisible imaginary have no real weight.

And, yet, the original and underlying tiny
And simple uncaused basis
Of all subsequent complexity
Could not have ever have been created,
For causes do not go back beneath causes forever,
And, so, there was a first 'something'
With nothing prior to it
To give it any design or direction.

Even if we endow Nothing
With the capability to differentiate itself
Into plus and minus,
Then that capability would
Have always been there,
And even if Nothing could do something
It would have always been there also.

So it is that nothing was ever created,
But that the original 'something',
Underlying all,
Went on from the simple

To the ever more complex,
That complexity being seen
Only at this end of the spectrum.

If there was no creation,
Then there was no Creator,
And all is still here that was ever here.

'Something' was the natural state of affairs
Since a total Nothing couldn't do anything at all.

So, one has it that man is separate from God,
But can get closer to God,
This being a kind of 'Heaven',

And that, while material, too,
Is separate from God,
God can mold it and affect it.

Another has it that we
Are the Dreamer and His dreams
Since there is no real 'we'
And no real material anything.

Yet another has it has it
That a Scientific Deity of sorts
Made the base material
With His aims already in mind and built into it.

Others might have it that we reincarnate
And go up a ladder, or down.

Austin has it that an Intelligence system
Would be a system
And so would not be fundamental.

Anyway, the 3-4 claims
Are many steps removed from reality,
Not just 2, for example,
The first two steps being
To ignore what is
And to then further replace it
With what isn't.

That is only the beginning
Of the many steps beyond
That build the house of cards.

As for an initial Intelligence for the universe,
What would it be doing

Just sitting there, all fully formed
With all its abilities to the nth degree,
Being just luckily there
In that fortunate position as the Main Man?

Only the question of 'Why Me'
Might have given Him pause,
For His talent wasn't even God-given,
It was ...?

And why did He take 13.5 billion years?

So, the single universal mind,
If one comes about,
Is in the future, not the past;
Some have been looking
In the wrong direction.

ALL IS AS REAL OR NOT
AS THE ADVAITA VEDANTA.

The Dream notion, if it be held,
Must be carried through and through.

If so, the old Indian dream writings
On dream paper cannot just be seen
As the only real and therefore
Meaningful kinds of ideas
When other dream writings
On dream papers are not,
For all would be but writ on water
With the feathery quill of smoke and fog.

UNGROUNDED BELIEFS UPHELD

A belief hangs in the air,
Because it is upheld by the owner of the belief.

A pyramid stands on the ground
And shows that a single top
Can be held in the air through support,
But the single top can itself not
Stand on the ground without support.

A DOT:

.

Now, this dot above,
Like the dot over an 'i',
Contains billions of atoms,
So it is really a big dot.
So, let's talk about a tiny dot,
Which we can't see.

It perhaps makes particles,
Like the quantum realm does.
Some are stable, some aren't.
It might make a virtual pair.
('Virtual' is a misnomer; they are still real).

Somehow, the members of the pair get separated,
They then going off to do other things.
The same with more.
They might also cancel each other right away
And go back into the dot.

The dot marches on...

Still, all began from a state of near zero energy,
The tiny dot all that stood between
Something and a total Nothing.

Gravity and all matter still cancel out,
But for the tiny dot.

Now, on average, the tiny dot didn't do much,
Many of the 'virtual' real pairs collapsing back in,

But, why did the dot do anything at all?

Well, it is that the simple is unstable,
In the sense that it always
Goes through phase changes,
And so did the simple stuff
That it made do so as well,
Then increasing in complexity
From quarks to protons
To stars to all the atomic elements
To molecules to cells to life to consciousness.

But how was there such a
Massive outpouring at first?
Well, most of the time there wasn't,
But probability goes through all states,

Even its far reaches of the very rare,
And so, once in a zillion years
A lot of stuff happened
Within the same time frame.

Either that or the tiny stuff emitted
Accumulated over a zillion years,
But, either way, a universe happened.

The dot tells all... the void nothing...

A void could not have an 'into',
For that is a where and a place,
And a void has no properties,
Such as extension,
For a Nothing is defined as not to be.
Only existence can be.
So, the void is nonexistent.

There couldn't be a Void, even for an instant,
Or nothing ever would have been here
(The meaning to be taken either way).

I'll give a little and refer to the Void directly
To say that there would have been nothing outside it
To maintain its perfect non state.

I'll even give a little more and say
That the bottommost lawless law,
Before the laws relating to the physical,
Is that anything but nothing goes,
And, so, the Void, being the simplest of all,
Follows the instability that we always see
At the simple levels
And so it must jiggle as the wiggling dot,
Albeit that the dot's plus and minus emissions
Cancel out if they remain in the vicinity.
It is, then, only the capability of the little dot
That truly is, in and of itself.

The accounting report of the Universe:

The conserved energies of the universe add to zero.
This includes charge and angular momentum, too.

We will forgive these sloppy accountants
For not being precise and leaving out the tiny dot,
For the dot is very small,
Compared, to, say, a dust mote,
Which contains half a million atoms;

However, that is no excuse.

The super accountants recognize the tiny dot
As being part of the equation.

All almost equals out,
The credits and debits,
But for the dot.

Go dot, go!

Everything has been compressed down
To two fundamental forces:
Yin/Yang from which all things
Sprang into existence;
One balanced entity
With two complementary
And opposing forces,
Positive and negative.

No outside...

What was outside of it? Nothing.
We call that nothing the void.

Yes, nothing was outside of it,
Not even a thing called a void.

However, let's just stick to the dot,
For, either way, the void is not a factor
In the doings of the dot.

I figure that electron/positron pairs
Come out of the dot more often that other pairs,
Since they are lighter,
But I guess all else comes out, too, doesn't it?

The Dot Proof:

Since the conserved energies
Of the universe sum to zero
And zero can't produce anything,
There must then be a very small dot
That made everything.

There couldn't have been
A whole bunch of stuff,
As the original basis
Since there was nothing prior
To determine that precise amount.

However, when positive
And negative energies cancel out,
But for the dot,
Then there is no more paradox
About the exact amount of the particle pairs.

The good old (ageless) dot made all,
In equal portions
Of plus and minus charge and energy,
Much like the quantum realm
Now still emits 'particles' in pairs.

The dot always was and ever will be.
It was the original causeless something
That had to be, for Nothing can't.

The dot was never created,
Having always been there.
Maybe there are even lots of dots
Doing the same thing.

It's really small and simple; however, from that,
The more and more complex occurs
In the ever increasing composites building
Up and up and up to the most glorious assemblies.

This popping in and out, at random,
Whether from a dot or a fluctuation
Or from anywhere that is the causeless
(Where the buck stops)
Is because there is nothing prior to it
To impose any order or design.

From then on, and upwards, though,
There are always causes
For all the subsequent effects.

So, we have to wonder why the causeless,
Be it the dot or whatever,
Didn't just stay in balance.
This would be an inert state.

Or, if, say, there was no more inward direction,
Making it bounce, why everything it emitted
Wasn't just so totally unstable
That it instantly decayed
And went back in,
This being as useless as it being inert.

Well, we know that there was
No complete inertness in our universe,
And we can't really talk about other universes.

We'll just have to guess,
So, my guess is that the causeless
Can have no real structure,
Although it could settle into one, given time;
But, generally, when there is no cause,
There would be some kind of 'law'
That the causeless can have no true form
And thus no inherent inertness,
That being a definite state requiring some design.

Another guess is that tiny simple things
Can always build into lumps,
Given that they would create
Or have some forces or gravity,
Although, still, not that these lumps
Would necessarily go any further.

Let us trace back the planet Earth:
Did it not begin solely by two dust motes
Colliding and clumping into one?

So, does causeless mean that something
Has to happen further of it?

PIERCINGS

Why don't people who wear jewelry
Realize how ridiculous they look?

I had a ring once,
But I fiddled with it so much
That I removed it.

I have enough holes in my head
Without adding more.

Jewelry, designer clothes, perfume,
Plastic trainers, luxury cars ...

Yes, they're not needed,
But I really have to keep my
Super-duper Apple Mac computer.

FOR NOW,

A total lack of any thing,
Including capability and possibility,
Would be a total Nothing.

If this were the case, it would still be the case,
As not anything can become of Nothing.
I note that Nothing has no properties;
No what, no who, no then, no when, and no where,
For it is not even 'there' to be able to do anything.

Obviously, though, there is 'something'—
From an uncaused ground-state capability,
And, so, this is normal and natural.

It seems that this ground-state
Must be tiny, a near 'nothing',
As it must be the residual
After all the conserved energies sum to 'zero'.

Since this ground-state
Always had to have been,
There was no creation of it,
And, thus, no Creator.

This argument even cuts out,
At the very source,
The possibility of a Deity,
Much less a Theity,
And we haven't even mentioned
That composite complexity falls
Way at the other end of the spectrum,
And that the causeless,
Having nothing prior to it,
Would not have any intended design to it.

NATURE

Never did I see a life so alive
As when the flowers thrived;
Now they're all gone and done,
Each and every last and single one,
But their leaves soaked up energy
Into the bulb for the spring to see.

SOMETHING IS REALLY "OUT THERE" SINCE

1.
Instruments can measure it.

2.
Starlight arrives from before humans existed.

3.
Things progress and operate in our absence.

4.
Consciousness observes experience.

What if all is just a projection?

It still comes down to the real at some point,
And what is of the real is real.

1.
Brahman dreams reality; yet He is real.

2.
Holographic projection;
The 2-D plate and the projector are real.

3.
A perfect virtual reality operating
The same as real; V=R;
A difference that makes no difference
Is no difference.

Reality is the message,
Regardless of the messenger
Of the mechanics of the implementation.

CHARGE

As electromagnetic beings,
We are going to carry a charge.

Females carry and use
A lot of charge cards
(And men, some, too).

Now, due to the economy,
These cards need the plastic surgery
Of being cut in half.

US MAMMALS

We mammals are not static;
We constantly learn more, make new associations
Discard old ones that failed, remember new facts,
Even constantly tweak our methods of analysis,
Thinking, writing, and imagining.
We can get outside help
From books and the internet,
Both an extension of mind.

We operate by what we have become
And have access to;
We wouldn't want it any other way,
Such as random events overriding
Our knowledge base and our selves personality.
We would then seem as air-heads.

Even Einstein didn't start from scratch without facts.
Prior physics informed him of the properties of light.
He knew that light was fixed and that it couldn't give,
So he imagined that perhaps time could give,
And so he tried making time a variable and it worked out.

For space-time, he noted an unusual coincidence
Of the rate of gravity and acceleration being the same,
And so he stood on the shoulders of giants
And made a new idea.
Now, maybe light bends for another reason,
Such as it is crowded with photons near stars
But he got the curvature idea going and it got measured.

Even when we see actions by others
Our mirror neurons mimic it,
If only to better understand the action.

People seem to use existing ideas
To make something new out of the combination,
As very little ekes out purely original
that is brand new under the sun.

As for how we seem to ourselves,
Note that our conscious state of being
Doesn't directly inform us
On the actual states of being beneath.

We see colors but feel nothing
Of the 3 types of proteins in the retinal cones
Rotating according to the amount
Of corresponding primary E/M 'color' coming in.

Oxytoxin helps bind the mother to the newborn
Until familiarity takes over.
Other bonding hormones promote friendship and love.
Senses, memory, and our ways of relating information
Present a unified picture to consciousness.

We don't feel the brain's systems operating
At their pre and mid stages.
It would be a mess of confusing data.
Anyway, it all breaks down to electronics,
Bio, and chemicals.

Consciousness may just be
An epiphenomenon of billions of neurons
Doing the true and actual work of analysis,
Only being able to inform us a few hundred milliseconds
After the neuronal analysis is nearly done,
Making us seem as mere tourists along for the ride,
But the global whole is then taken back in
And incorporated pretty much instantly
And so it can influence immediate
And further analysis and action and so on

LEAKS

On another random topic,
My IBM office mate had been designing
A floating point instruction
That had to follow some random rounding
Or something like that.

I think he had to make a history
Of the randomness to keep it 50/50,
But I could see that he was sweating
This waste and some other things.

I asked him how it was going.

His immortal reply was
"Everything leaks."

A VERY CALCULATING ACCOUNT OF LOVE

Professor ProfPat, having taught Accounting 101.0,
Once gave these opening remarks while mindfully
Undressing the entire female Catholic student body
And repeated the points here:

We will use the calculator to count the numerous ways
That love can be made, even times 2,
As well as to use our many calculating buttons
To more fruitfully multiply.

As for the zero point, there is no dividing that way—
So always wear a raincoat while in the shower,
For if the Maker wanted us to be totally naked,
We would have been born that way...

Oh, never mind, but please do wear a tiny raincoat
With a little hat to safely have sex
And therefore, of course, more oftenly.

Plus and minus can make
An overlapping summation of one,
Or it can amount to a big 0—
Hopefully with a lot of steam and smoke
Made by fire and water.
If you smoke after sex,
Then you are obviously
Doing it much too fast.

Never place a square root in a round hole.

Be wary of crediting a debit while in the red.

Use the double entry system.

Do not become a numerator
If the sign of the denominator is
'Do not enter'.

Numerators and denominators may be reversed
In cross multiplication for those indivisible.

Love divided by infinity still equals love.

Class over.

DETERMINISM

I had an insight into determinism,
Although I somewhat lost my train of thought
On one good sentence that fled.

If all is determined,
Then this would have been the case
Throughout all history
And on all the way back
To the original basis of all,
Although the initial state
Could have been arbitrary.

There would have then
Been a definite boundary.

Before that (if any)
Would have been nonexistence,
(Here I had something else in mind
But can't remember, so I'll say...)
And yet nonexistence
Cannot be 'there' in any way,
So I reason that the original 'something',
Whatever it be, had to always have been around
As the natural state.

However, that logic above is only half the story.
Given that the beginning of secondary existence
Was probably arbitrary, seeing as the uncaused
Would have no definite plan imposed on it
That was determined.

Is what it did produce
Still determined all the way up from there?

And, if not arbitrary,
But from some 'law' of nature
Due to whatever constraints,
Still, is what it produced determined
All the way up from there?

If there are random quantum effects,
Wouldn't they cancel out,
Still leaving a lot of atoms to form things?

Are not some structures pretty much immune
To stray atoms coming and going,
At least for quite a while,

Like a star or a car?

But, what about, say,
Atoms inside a structure or a human body?
Well, they come and go, too;
However; all atoms of a type are the same,
So, this doesn't matter.

Then what of the atomic arrangements
That we do have,
Plus the molecules that make cells?
Are they true to their structure
In the way we remember,
Imagine, think, and do?

Every time we measure a quantum item
We get a different answer for some properties,
Although not its size.

So maybe the result of a bunch of atoms
Bumping together is not for sure,
But perhaps once molecules group
And go further into making cells
They then operate as they must,
Outside of aging and diseases happening?

For the most part,
I believe that we would want our thoughts
And actions to be dependent
On what we have personally become,
rather than on some random dice throws,
Except maybe in cases
Where we accept 50/50, due to a tie
Or the final not mattering
Since either way is acceptable...

Our will depends on who we are,
Our learnings, associations, and memories.
This will of ours is not free to ignore all that,
Is it, even if it seems to be?

It would seem not to be.
Will is not fixed though, for we learn.

When ProfPat was agnostic,
That was his will back then.
He then incorporated influences
And so his present will is to not be agnostic.

He wouldn't want his present outlook

To suddenly flip for no reason;
He would want his will to follow
The logic and reasoning that is him.

None of this means that the will doesn't use
All of the brain's analysis,
Just that it is restricted to it.

We wouldn't want anyone else thinking for us,
Except maybe when we have to take an exam
That we haven't studied for, in in a sports contest.
Ha-ha.

...So, is determinism really so bad,
Seeing as its opposite is 'undetermined?

Would our will be free if will was undetermined?
Well, yes, but more like crazy,
Mental, random, and air-headed.

So, if will is determined
By what we have become,
Are we robots?
Yes, our learning programs us,
Although, like some present smart robots,
We can absorb new information
If we are open to it.

If not, then we are stuck.

Was human type life
Determined to arise in this universe,
Given all the planets
And enough time for evolution to work
With all the material?

Well, we do have one case,
But what if asteroids
Didn't wipe out the dinosaurs
Before time ran out?
Or was this determined
By the initial arrangement of the matter?

I'd say not,
But perhaps some
Other kind of extinction
Would have arrived sooner or later,
Although maybe not.

A clearer case is that
Of the weather and the climate,
For, while it could change any time,
The exact times don't matter so much
As that it keeps on varying anyway.

What about just any kind of life?

CONSERVED ENERGIES

Scientists say that the conserved energies
Of the universe sum to zero,
Including charge and angular momentum,
Even that the negative potential energy of gravity
Equals the positive kinetic energy of matter.

If all ever canceled out,
What's left would be the sheer capability
Of the plus-minus division that started it all,
Something probably very small
Since it is only the remnant
Of what sums (almost) to zero.

If it were really zero,
Then nothing would be there.

Is it the 'dot', the tao, seed substance,
A quantum fluctuation?

It doesn't matter,
For the fundamental cause
Couldn't be Mind Aware,
For a complex composite's parts
Must first be there.

What about from a Void?
Well, a 'void' is void of everything;
It has no what or where.

NUTTIN'

Ain't nuttin' that can come from Nothing—
Since Nothing ain't even there to become anything;
However, even the 'threat' of this perfect
And unstable state may cause some jiggling to happen

TOE

Conserved energies sum to near zero,
'Near' because 'zero', a void of Nothing,
Is void of any properties.

What is left over is
The necessarily tiny ground-state.

In the name of Occam's Razor,
Nothing is the simplest state, in principal,
But in practice it can't do anything,
Much less have a where of being there,
So then the next possible simplest state
Is the little old near nothing
That is actually something,
But not much to speak of.

Now, at the complexity end of the spectrum,
Where we are, there is much to speak of.

'I'

The raw scene in the eye
Becomes seen by the visual system,
That lovely painted face evolved
For the 'I' to know.

The scene seen is eye to 'I'.

The individual "I's",
Each unique in all the universe,
Think somewhat differently,
And separately,
Due to their individual history
Of nature and nurture—
And even diverging more
Than that due to the indeterminate chaos
Of the quantum jitter bugs
Dancing that which ever had to be.

The brain "I's" the scene
Seen by the eyes.

BRAHMAN

Brahman was [automagically] always around,
And very sleepy, I guess, and so He had a dream.
Do you expect anyone to believe this?

What does he do when he is awake?

Perhaps I am not being clear about
The direct questions
Of the contradictory ego principle,
Such as having to ban it.

So how come we have ego,
As Brahman certainly does,
Us being in his image,
If then not to use any of it?

Was the guy dreaming
A nonsensical dream
Or even a nightmare?

Or do you really mean that
He made real actual stuff
Instead of the stuff
That dreams are made of?

How do characters in a dream
Take on a life of their own
And decide things like eliminating ego?

Maybe the Dreamer should WAKE UP,
For he is a dream of an even higher power perhaps.
Does he ever wonder how he got the job?
Right place, right time?

If you, a separate figment
Of Brahman's glorious dream imaginations,
Have any dispute with his other separate figments,
You must take it up with him (Brahman),
For it is his playground
And Him talking to Himself.

That is, if you can ever stop him from snoozing.

THE THEORY OF EVERYTHING

The TOE is that now of the several toes,
Plus the toe food for thought zinging in between
That is the electron, where water flows
To keep the toes working well and clean.

WORLD CREATING

This capacity, which in our own species
Is great enough to be termed "world-creating"
Is present to a limited extent in the higher animals,
And less and less in the lower ones,
As we go down the ladder.

So it is that man can be called "world-creating",
In comparison to animals,
Who range from "poor in world"
(the higher ones)
Down to nearly worldless
(the lower)

And, something like a stone
Can be called "worldless",
Period, since the stone does not care
In what situation it exists,
And any time horizon of its being
Is going to be a very long one,
And again of no concern to the stone.

The other respect in which we are world-creating
Is in our acquisition of newer worlds,
In which we have far more power
Over our environment,
Replacing older ones,
In which we are at the mercy
Of witches, wizards etc.,
Who we must try to pacify...

This is why Heidegger viewed
The development of technology
As a question of
Our "enframing" our world differently
Than before, surrounding ourselves
With a very different scheme of entities,
To which we stand in a very different relation,
Than the wizards and gods of before.

FELT SENSATIONS

We have feelings, thoughts, and sensations
That are presented in consciousness,
These being the brain's perception of itself,
Of its doings from analysis of inputs
And what we have become in our selves.

This state of being that is observed
In consciousness is a reasonable painting
Of a face on the neurological states beneath,
Which, absent any disease of the mind
Affecting its mechanics,
More or less faithfully represents
One's learning, inclinations, associations,
Personality, and memory stored there.

The other 95% of what the brain does
To maintain itself and the person
Does not surface into consciousness,
And need not,
For that could be a mass of confusion.

As an aside, we see that consciousness
Is a tail-end process,
The brain's binding of information globally,
But this does not mean that
We are merely tourists along for the ride,
Because the global information
Is still remembered by the brain and its will,
Like all that you are,
Still influencing future thoughts and actions,

Distancing quiets the neurological area
Responsible for knowing where your body ends,
Thus, the boundary appears to be no more,
As one would expect, seeming as if
You have merged with the cosmos.

This is still feeling and sensation—
And it is only imagination that wishes
To make Something More of it,
A "more" that is quite extensive
In its implications, I might add,
And thus a "leap" of inference.

A quietus also overcomes
The neurological area responsible
For the identification of the self,
Which then gives the feeling

And the sensation
Of the self disappearing.

Thus, again, this painted face
Is presented to consciousness,
Just doing as it should.

Imagination again tends
To make Something Else of it;
However, it is natural that
The notion of the self diminishes
When one quiets that area.

To get around these conclusions,
One would be forced to extremes
To claim that the known organ
Of the brain does not exist,
Or that if it does,
That it doesn't do anything,
Replacing it with the unknown
And necessarily invisible soul,
The "leap", which then "allows"
Even further leaps to the notion
Of the supernatural and then onto God,
Which was already the desired end
In the first place,
The (in)vested interest,
For others reasons
As best explained elsewhere on TQ.

"BUT ALSO THOSE WHO LABOR ON THE SABBATH"

At a hotel I saw a 'Sabatt' sign on an elevator.

I found that it stopped on every floor so that
Those observing the Sabbath could then use it
Without the 'labor' of even pressing a button.

Up in their rooms,
They could 'accidentally' flip a light switch
By just 'happening' to rub against it.

MISSING LINKS?

Of the Ground-state of All?

The link is not missing.

The ground-state existed always
Since not anything could have become of Nothing.

Missing Links?

Attempts to fill, via Imagination

Imagination/Mind wants an answer;
That is its job.

It may not compute,
But must know,
But it can't,
But it must...

The upper floor,
Especially the holistic right brain,
Can entertain imaginations,
Some of which wonderland
May be ungrounded in reality or detail.
One can get grooved.

Why not intercession of left brain?

Heightened right brain activity,
Or one not used to left brain;
Not mechanically inclined
Or encultured,
Or busy going about life.

Imaginal idea has no detail.
Feels good. matches wishes,
Creates stopping point,
Case closed.

Meaningful because it was "thought of"?
Rather, thought came, perhaps of vested interest.
Difficult to ignore one's own thought.

Can see/use, say, a car as a whole.
May drive too fast with low oil or air.
Zen and the art of...

What if details very much requested by others?

Neglect;
Repeat imaginational idea;
Employ magic;
Go to extremes;
Brain does nothing;
Reality is a dream.
Evolution never happened;
Imagine the soul to replace all of nature.

More imaginal ideas added on for support,
Such as God, rules, etc.
Holy-holy aspect glued on
To neutral energy and material.
Get mad to show that
One's imagination is irrefutable.

Left/right brain synergy:
The detail up and the whole down
Can meet in the middle;
Better understanding of
Each direction and overall.
Rounded life equals
The blend of yin and yang.

Fix?

Difficult to impossible now;
Maybe in next generation.

Ex: Demotion/promotion Idea:

A certain dog or a cat may actually
Be a person who did bad in a prior life.

Why pick example afar from most?

Local imagination ideas too close to home.

Go into their shoes;
Non-reincarnation beliefs seem just as weird.
However, details of mechanism absent.
Dog brain or human brain trapped in dog?

Conflict with contrary ideas elsewhere
Means difference = seems really strange =
Protect idea and minimize the contrary =
Evil [sometimes].

Purely imaginary concepts
Are a flawed concept of 'right' and 'good',
Yet are not even told as theory,
But said and pronounced as 'is'.

Indoctrination of young.
Impoverished need help from the 'outside'.
Myths grow.
Stable countries find help at hand.
Myths diminish.

Same Story.

States of being on upper floor are 'directly' felt;
States beneath are not, or not known of,
Or don't seem to pertain.

Why?

We are not privy to occurrences beneath.
Brain info is relatively new.
Not all have education.
Not all wish to research and learn.
Cells, molecules, atoms seem lifeless.

We want what we want.
Scenarios of consequences collapse
According to what one has become
However, feels natural and instant.

Idea: Meditation = Ultimate

Still felt sensation
[Of minimal neuronal network],
But it yet intercedes, as must.
Not really on ground floor;
Can't get there;
Feeling is still sensation on upper floor
Of not much happening and not much self.
Seems one with Cosmos.
Minimal boundaries.

The other side: Too much left brain.

Too much into detail.
Can't see the forest.
OCD.

Useful use of imagination:

Grounded or semi-grounded imagination
Leads to anticipation and then feedback
Of the subsequent events.
Imagination is then refined, added or subtracted.
Continue on with actions.
Reimagine...

Totally ungrounded imagination
Has no subsequent events,
For all is invisible.

The end.

Ignore what is to make up what isn't.

Double error.

THE PREFRONTAL LOBES
JUST RECENTLY APPEARED?

I don't know when frontal lobotomies began,
But those operated on,
Aside from perhaps being
A little 'off' to begin with,
Lost whatever planning abilities
There were in the lobes,
So, maybe the use of the lobes
Was in happening long before then.

Hope our lobes continue
To get used more and more,
For it is the lack of planning,
That is, consequences unforeseen,
That gets many impulsive kids into trouble.

They ought to dwell on these instances in schools;
We could drop some parts of math teaching
That never get used.
(There would be the option of taking higher,
Complicated math classes for
The scientifically inclined.)

THE FRAMES THAT WORK

There is no time for space, no space for time,
They being of a separate reason and rhyme.
Time is a then, now, or a when of what is placed
And replaced in that place that is spaced.

At that long gone doorway
Of where the origin once was
I knocked onto the thin air,
For there was no one there.

The thought had arrived, unexpectedly,
So how could s/he will
What was already here.
Whence did it bud, so silent,
Before it grew to fruition?

The egg falls and breaks
From heights much too high,
But on the ground of being it rolls along;
The good eggs wobble, taking the dips and dales,
For within the shell grows the fertile embryo.

Fundamental parts compose
The physical and mental parts
Of the systems above
That are made of those basics below—
The complex composite movement
Of being and mind.

Awareness, thinking, doing, dreaming,
And seeing takes a "village" of constituents,
Those always having to be first.
We cannot just wave a wand to make it not so,
Absolving God from it by exemption.

The nervous system connects to the brain;
When you feel a pain it is known in the brain,
Relating the sensation to whence it came,
So, it is all but an extension of the brain.

DIA ADVENTURES

Before the mission, Passiona and Austino
Had driven to the end of Kuhio Highway 56,
Reaching the exotic Ha'ena State Park
Located on the north shore
Of the Hawaiian island of Kaua'i
Often referred to as the 'end of the road'...
However, it was their beginning as a married couple.

They were tucked against the Napali cliffs
Is this Ha'ena State Park.

"Ha'ena" is usually translated as 'red hot'.

When the sun is down on the
Right side of the Napali cliffs,
The scene turns to a deep and perfect red,
And thus is where many couples
Have envisioned a beach wedding.

The 230 acres park is situated
At the terminus of the North Shore drive
And is host to Lumahai beach, Ha'ena beach,
Ke'e Beach, and a spectacular
1,280 ft cliff named Bali Ha'i.

The cliff and these beaches
Have also been the locations
For several well-known songs
In the 1958 film titled Bali Ha'i,
Set in the South Pacific by Hollywood in 1958.

One mile to the east is Lumahai Beach,
Which is actually three beaches
In various degrees of connectedness,
Depending on how the sand builds up.

It is visually stunning, with black lava cliffs,
White sand, blue ocean and green jungle.
It's always great for running on soft sand,
Then swimming in the fresh water
Of the mouth of Lumahai River,
And playing in the waves
Where the river meets the ocean.

And now they were in
The clutches of the enemy,
Perhaps even intentionally.

CONCEPT

Define a "concept" as anything distinguishable
From any other concept, including nothing.

Define reality as the set
Of all possible distinguishable concepts.
No problem, it's countable.

If reality were such that
No concepts except itself were possible,
Clearly no universe nor any Creator could exist.
Luckily for us, that is the case.

Consider only primary concepts,
Excluding those that appear
Only within other concepts within reality,
Such as a mind.

We can assume reality is timeless
Since a concept that is possible at any time
Is ultimately possible from the beginning.

Being timeless, it is eternal and unchanging.

What never exists cannot ultimately be possible,
So every possible concept must exist.

Since reality is timeless,
Whatever actually exists cannot alter reality.
(It can be considered a synonym
For "nothing" if you wish.)

This is possible if and only if it is not observable.
Any information that approaches
Observability by reality
Is necessarily canceled by its negation.

Uncertainty and increasing entropy
Hide information.
It isn't clear that energy is relevant.

Actually it is not just luck
That makes concepts possible.
The nothing we are talking about
Requires a preexisting prohibition.

The reason there is no Allah-God-whatever
Is that it is undefined unless it is either:
The set of all possible worlds,

Which requires no entity at all,
Or an arbitrary invisible ruler of one planet,
Which makes no sense.

The universe is here now,
Which is not a special time,
And so then a universe
Could be anywhere and whenever,
There even being more than one at a time.

Reality is timeless,
Being eternal and unchanging,
Since a concept that is possible at any time
Is ultimately possible from the 'beginning'.

AWARENESS

The 'I' of awareness,
Which is the same
As the normal usage of 'I' in English,
Witnesses, in consciousness,
What surfaces from the self (the brain)
Of thoughts, feelings, and sensations
Associated with the person's
Memories, learnings, etc.

So, 'I' feel [whatever]
That surfaces from the brain's analysis
That is globally bound
Into an experience in consciousness.

While consciousness is the last
To be informed of the events
Amounting to the experience,
It is not the end of the road,
For that experience, too,
Becomes part of you
After only 200-300 milliseconds,
And so you can easily
And quickly use it in your next idea.

"'I' feel happy",
For that's the message arriving now.

'I' also feel that I have to
Go clean the basement floor,
For my senses noted
That it's full of dirty footprints, etc.,
And on up to consciousness.

"FELT"

"Felt" equals sensation,
For that is a part of what the mind does.

There are many substrates beneath,
Even before getting to the level of atoms.

Imagination then takes the sensation
And flies off with it to pronounce
And declare meanings upon it.

The soul is.

Not 'is', but just an ungrounded theory
Of the invisible preached as fact,
A deception, an enculturing born
Of the Trickster right brain
That neglects the details of the left.

What you see and feel within
Is the brain perceiving itself,
Through a system built through layers of neurons,
Nerves, and cells, on down through molecules
To atoms to electrons and further.

The imagination would have to
Dispense with all this
And throw it out the window
In order to replace its doings
With something invisible;
But, then again, some have stated
That the brain does nothing
And so these are the lengths and extremes
That they must go to, ignoring what is
To replace it with what isn't;
But, facts always trump no facts whatsoever.

It is also way off to claim provocation
Because a side of the discussion
Is contrary to one's own stance.
You are on ToeQuest, are you not,
In a thread debating the soul?

Stress is the difference
Of what you expect to happen (wrongly)
And what actually happens in reality.
Nevertheless, detachment is a method
That anyone can employ.

When you transcend biology you find your soul!!

You would have to show that this theory
Of there being something beyond biology
Is not just the looking into your own brain.

Until then, it is not proved
And so it is just a theory—a notion.
Same with extending the theory
To say that it is from God
Same with then extending it
To say who God is and what he wants, etc.

LIVING OFF OF THE GRID

The motorcycle churns the dirt of the trail,
Its first gear pulling up and up,
Through twists and turns,
Over roots and rocks,
Towards the camp,
In late afternoon (raising up the sun),
Sometimes even at night,
Water and goods in its saddlebags.

Here the tent, the soft moving airs,
And the lightness of being;
Here the internet from the fort to the laptop
On a small folding table,
Then, later on, the soft pillows of sleep
Into the dawn below the shooting stars;
Existence always trumps essence.

Here today, gone to-maui—to relearn hang-gliding.

THE HUMAN MAGNETIC CYCLE:

This collects coins that fall out of your clothes
In the washer or dryer.

It's also a kind of personal transport
Based of the principles of the monorail,
Plus, a monthly biorhythm
Based on the earth's core

OBES AND NDES

NDE tunnels of light and such
Can be explained by neurology,
And OBEs by a condition called sleep paralysis.

They can also be chemically induced,
Resulting in full blown episodes.

Neither, then, are proof of a beyond,
But of an altered brain state.

I had several OBEs.

In the first one,
I noted that the scene
Looked as real as real could be,
But I did nothing further
Than to float around the bedroom,
Full of amazement.

I figured that the dream model of reality
Is the same one that is employed
When we are awake.

During the second OBE,
I rearranged the items on my end table,
Even knocking one item off.

All still felt totally real to the touch and all that
And I was sure that I would see the evidence
Of the end table results later when I fully awoke;
But when I really awoke
I saw that nothing had been moved.

I also found that I could awake
From dreams anytime
By clenching my whole body,
And so during the third OBE
I luckily found myself in a kind of halfway state
In which my dream-arms
Were seen to be fiddling with the end table stuff
While I could also see my real arms
Yet lying beside me, unmoving.

Another time,
I kept dream music playing after I awoke.

I guess the moral is that
Sometimes a virtual dream reality

Cannot be told apart from real.

I was so sure that I was out of my body,
But one must also remember
That memory and imagination
Often images scenes from above (try it).

It is also the case that people of different religions
See different religious symbols during NDE's,
An indication that the phenomenon
Occurs within the mind, not without.

OBE's are easily induced by drugs.
The fact that there are receptor sites in the brain
For such artificially produced chemicals means
That there are naturally produced chemicals
In the brain that,
Under certain circumstances
(The stress of an trauma
Or an accident, for example),
Can induce any or all of the experiences
Typically associated with an NDE or OBE.

NDE's are then nothing more than wild trips
Induced by the trauma of almost dying.

Lack of oxygen produces increased activity
Though disinhibition—
Mental modes that give rise to consciousness.

What about the experience of a tunnel in an NDE?
Well, the visual cortex is on the back of the brain
Where information from the retina is processed.

Lack of oxygen, plus drugs generated,
Can interfere with the normal rate
Of firing by nerve cells in this area.

When this occurs 'stripes' of neuronal activity
Move across the visual cortex,
Which is interpreted by the brain
As concentric rings or spirals.
These spirals may be 'seen' as a tunnel.

EPITAPH FOR SYMBOLIC IDEAS
OF THE BIBLE'S CONTENTS

Evolution anywhere could start
As soon as there was a place,
Which probably wasn't there
At the beginning of the universe.

Life on other planets is indeed probable;
Perhaps a microbe from one landed on Earth.

Evolution clashes with
The Designer for many reasons,
One of which is that
It doesn't require the Designer,
For it happens via climate,
Weather, and the environment,
With only death as the chooser
Of the pointed from the pointless.

It was a design without a designer,
And thus it took a very long time,
Much longer than a week.

It also shows the opposite
Of the immutable forms
That are claimed by the Bible;

Not that the Bible has any credibility,
For it also got cosmology wrong,
As well as having no revelations about the future
That a human could not have possibly known.

So it is that if the Bible gets the natural wrong
It can hardly be depended on for the supernatural.

Yet, a clash remains over what was written in the Bible,
Which is no wonder since it was written by man,
Differing from what was actually found to be the case.

The Old Testament was of some old Jewish legends
Of the many gods amalgamated and rolled into one.

The fundamentalists couldn't stray from the literal,
Making the Earth but a few thousands years old,
But some astute Christians realized the folly,
Knowing that without the Bible all would be lost;

So they merely declared and pronounced
That the Book for the common man,
Written in plain language,
As had long been advertised,
Was now full of secret symbols
That only the experts could decipher!

The Bible had risen from its own ashes.

DIA...

Rascal placed a call to the head of Mossad,
Which is the Institute
For Intelligence and Special Operations,
Being the national intelligence agency of Israel.

"It seems that Iranian scientists are disappearing,
That much is going wrong in their nuclear plants,
And that many of the needed shipments are not arriving.
Do you know anything about that?"

"Who, me?
No, nothing, but we know even less than nothing
About some other problems that they are having.
Do you know anything about those?"

"No, not a thing."

"Thanks;
Now we both know something."

"Good work."

"You, too."

THE SHIMMERING RAINBOW OF NOW

The light spreads forth all,
Illuminating reality,
Then gravity recalls it,
Erasing it.

(Written in the spirit of Nobody Nowhere)

CONSCIOUSNESS IN OTHER CREATURES

While there's no direct evidence
For what it's like to be another creature,
We can observe their actions.

Perhaps we might note a snail moving about
In some way that seems
That it knows little else but food,
And/or light and dark,
And/or warm or cold.
That might well be the extent
Of its smudge of consciousness.

CHURCH WOMEN

Women were treated very poorly by the Church,
Yet another not so fine move
That became of the Church's flawed concept of good
That included man as being superior to woman

Even Sarah Palin as Vice-President
Would have to be second to her husband
At home and in the church.

Humans are what they are,
No matter what the Church
Thinks they ought to be.

The actual origin of Man differs
From what the Church thinks.

Maybe they also think
That deadly viruses are from the Devil,
But, who knows anymore...

There is liability for both acting
And not acting to what is;
Dogma has trouble adapting,
And so it cannot act.

.

WORTHLESS THEORIES

A dead giveaway that a theory has a problem
Is when it neglects or is immune to any details,
Or when it magically amends itself to any turn of events;
These indefensible theories are not even interesting.

When pressed for detail,
These theories do not
So much address the detail
As to 'respond' to it
With completely
Off-the-wall type of answers,
such as:

— *The brain is not needed.*
For all is done via a link
To an invisible man in the sky.

— *All is a dream, so details are blah-blah;*
We don't exist.

— *An old book said that forms are immutable,*
Thus, evolution never happened.

— *Don't even ask, for all ego must be removed.*

— *Consciousness is no longer a subject*
Of experience, but is now an object
That floats around granting experience.

— *We come back as other creatures.*

— *We will burn if our free will*
Disagrees with some Other Will,

— And so forth, ad nauseum,
The 'responses', obviously,
Even greatly disagreeing with each other.

Thus, the right brain runs away with itself.

USES OF THE MOON

If there were no tides and tidal marshes
Filling and emptying, then what?
Would life have formed?

Isn't the moon really a planet?
Yes, it's orbit is everywhere concave to the sun
And is captured by the sun and not by the earth;
Together they are a double planet system
Whose center of gravity is still
Within the earth somewhere.

What about lunar cycles
And bio- or menstrual cycles,
Moon gods, honeymoons, romantic walks,
A base for space travel,
'To the moon, Alice',
And lunatics?

The reading I referenced
Was making the hypothesis
That tides created by a Moon
Orbiting at 30,000 Km from Earth
Had been "critical" for violently mixing oceans.
(Tides were far more great with a
So near orbiting object)

This mixing, lasted for at least 1 billion year,
Eroded a large amount of material
From new born continents
And this material continuously
Mixed and mixed again
Was at the origin of the first life,
Half a billion years after Moon formation.

No Moon, no party ... at least no party for life.

If this would be true,
The Earth collision and Moon formation
Would be the main first engine of life
And the probability of such an astronomical event,
I think, is very low.

The coincidence of this low probability,
Combined with the "improbable hill" life climbed
In few billions of years,
With the Fermi paradox ... is well ... amazing.

If life is so improbable,
Compared to the number of possible stars,
So strictly related to an almost impossible "collision",
No surprise we are not able to get any signal
From any nearby partner.

In the same reading I found the statement
That Earth's rotational axes will start to oscillate
At that time and that may will end
With Earth rotational axes in the orbital plane.

But for these accidents,
Including the elimination of the dinosaurs
And the fusing of a 'chimp' chromosome,
We might not have been here ever,
But as you say there are
Many stars and planets.

Moon Children

The Earth would wobble like a dying top very soon,
Without the steadying influence of our lovely moon;
But, it's slipping from our grasp an inch & a half a year.
The end's not so near, but we'll need a way out of here.

Yep. It is going away at a rate of 3.8 cm/ sec
(Thanks to the Apollo mission for this data).
I've been reading that Moon will leave Earth orbit
When it will reach the average distance of 440,000 Km

So in about 1 billion year 40,000 Km/ 3.8 cm*year^-1
(current distance should be around 400,000 Km)
The Moon will feel more attraction
From Jupiter and will say goodbye to Earth.

THE FOUR ELEMENTS—AND DUST

From the fires of stars to those of cremation,
We have breathed, flourished, and dissolved:
Life is ashes to ashes, stardust to stardust.

Of airy winds, vapors, and a soft earth,
We rest, at last, under the spinning skies,
Those of Earth's sunny days and starry nights

THE YEAR/DECADE
OF LAME MYTHS IN REVIEW:

1.
Evolution never happened;

Wait, unless it did and was God's plan.

2.
All is a dream or a holographic projection.

Oh wait, Brahman is real and/or the projector
And 2-D plate are real, so what is of the real is real.

3.
OBEs/NDEs are of real happenings.

Wait, one can only 'see' the top layer of the mind.

4.
Using numerology on the way a word is spelled
Can tell us something.

Oh, wait, words were not formed in the first place
With this purpose in mind.

5.
The brain does nothing.

Oh, wait, we can scan it and see it in action.

6.
A total Nothing can produce something.

Nope, for it's not even there to do anything.

7.
The invisible, unseen, unknown is there.

Oops, it has no effects.
We see only the natural,
Nothing extra-, beyond-, or super-.

8
'=' is the Toe.

Not; there are various forms,
From mice to galaxies.

9.
CO2 levels are not rising.

False, they are.
CO2 drains very slowly;
More is arising than can drain,
So the bathtub overflows.

10.
The right or the left brain is better than the other.

Nope, both sides are required.

11.
People wrote the Bible as a code.

Well, people did write the Bible
And that's what's wrong with it.
'Code' is mere interpretation as desired.

12.
Right brain people reply to detail in left brain posts.

Nope, they 'neglect' it.

13.
Ego is bad.

Nope, all things are only good or bad
In how they are utilized.

14.
*Conspiracies are kept secret
By the hundreds of thousands
Of people involved in them.*

Nope, impossible.

~~Myth~~

PARADOX RESOLVED

I used to think that a specific amount
Of energy/mass in the universe
Was a kind of paradox since
What could have defined
That specific amount
With nothing prior to
The uncaused ground state.

However, there is no real 'amount'
Since the conserved energies sum to 'zero', almost,
But for the capability to break the balance
Into any amount of pluses and minuses.

Perhaps a total Nothing
Was so perfect and unstable
That it had to jiggle about,
But, I note that Nothing doesn't
Even have a place of a 'where', nor any property,
And so I have to agree with Victor's book
That this something/capability of the ground-state
Was naturally always around, and still is,
Such as in the quantum realm
That emits in pairs of plus and minus;
This is the 'how' and the 'what'.

Since there was probably
Extreme inflation at one time
And even now the universe is expanding,
It kind of makes sense
That the large came from the tiny,
Its jitters now writ very large
As the galaxies or at least the CMBR.

What about the paradox
Of a specific 'where' for the ground-state?
It must be anywhere and everywhere.

Perhaps the stuff usually falls back in,
But every zillion years it takes off.

The Who?
None, as it was causeless,
Although we are now a 'who'.

Why?
The closest we can say is that Nothing cannot be.

When?
Any time, I guess, according to some probability

Any clear definition of the ground-state or what it can do?
Not really, as there was no state before it
For it to be specifically defined.

Also, about spacetime,
E/M is a self-regenerating wave
And so it needs no ether as a medium.

How far does 'space' reach?
Wherever the E/M
And other influences are and extend to.

Influences are made and are everywhere?
Maybe, maybe not.

What if we stick our hand beyond the edge of space?
Well, then we just create more 'space'
From our influences.

What's really beyond the edge if there is an edge?
Not anything. It's just not there in any way.

THE LARGE FROM THE SMALL

It has been said that galaxies
Owe their origin to quantum jitters
Suffusing space, enlarged.

Galaxies are nothing
But quantum mechanics
WRIT LARGE across the sky.

It gives me shivers, butterflies,
The willies, the creeps, collywobbles,
The heebie-jeebies, jitteriness,
And the jim-jams just to think about it.

THE COSMIC SUBWAY LINE

While the universe has no potential bounds,
Perhaps the largest a local object can be
Is just before it collapses into a black hole;
The smallest would be of the Planck scale;
So, then, black holes take in the energy
And transfer it back to within the Planck space.

MEMORY EXPLAINED

The past is never past,
At least while we are alive;
Our memories, though volatile,
Being both ephemeral and re-cognized,
No doubt have some basic persistence.

But how does this past remain,
And what kind of substance
Could there be
That lives outside time?

What makes it so strong
That it can survive
The merciless climate of the brain?

And in what storm's eye does it reside
In the center of the maelstrom
Of the change and growth of cells?

What be this grain that persists
Among the shifting sands of time?

All this we shall show
And answer soon
In the search for lost time.

Our rememberings try to describe
Reality as it really was experienced,
But, that sheer essence may elude,
Although some general outline remains.

Then, too, we add to it, subtract from it
And reconnect by association to the new.

Lo, the subjective metes out our reality;
While the objective lies furthest removed.

Perhaps, we may have a memory
That returns from a taste of butterscotch
From which Grandma's home then arises,
And then of connections further becoming.

How do some crumbs, here, and of the past
Waft back as vapours unto our present?

Do the senses of smell and taste,
Yet more fragile and more insubstantial,
Bear a unique burden of memory,

As more enduring and faithful,
Rising up past the ruins of the rest?

Just noting the butterscotch,
Back then,
Without its tasting,
Would not have made the mark.

Everything is connected
Within the mind,
Each germ of recollection
Ballooning into a revelation.

Time mutates some ancient pastimes,
And so they are not wholly recaptured,
And sometimes rather fallible,
Even altered more by the call to mind,
Yet they are there.

A memory begins as a changing
Connection between two neurons;
The strength of the synapse changes
So that the neurons can communicate.
Thus, the taste of memory
Also activates
The neurons downstream
To do with one's childhood days.

The neurons have been
Inextricably entwined,
Yet, too, reconsolidate upon recall.

How do we remember
Long after we have forgotten?
How do such apparitions reappear,
Some with no suggestion of their origin,
And sink and swell, float and change,
Withering the acids of time's reflux?

The memory making process need proteins
For the cellular construction of remembrance,
Yet the life of a protein is but 14 days.

And some hippocampal neurons die,
And some are born anew,
Yet some memory seems immutable.

Does the mind constantly reincarnate?

Aye, our memories must be made
Of a material stronger than cells,
And must be quite specific as well.

While each neuron has but a single nucleus,
It has a teeming mass of dendritic branches,
Connecting to other neurons
At dendritic synapses,
Such as the branches of two trees
Touching in a forest.

So, it is at these tiny crossings
That memories are made.
Not in the trunk of the neuronal tree,
But in its sprawling canopy.

What marks a specific branch
As a memory?
What molecule awaits
The taste of butterscotch?

It has to turn on mRNA
To help make the proteins.

It's name is
Cyptoplasmic polyadenenylaton
Element binding protein,
A tough assignment of a name
For even my memory to recall,
So, how about CPEB, for short.

Since it was in the brain's memory center,
Scientists looked for it in sea slugs,
Amazingly finding it in the slug's neurons.

Upon removing it, the sea slugs
Could not remember a darn thing!

But how does it work,
Existing outside of time?

Well, it has a series of repetitions
In its amino acid repetitions:

QQQLQQQQQQBQLQQQQ,

Where Q is glutamine.

Looking for similar odd repetitions,
What looked like a prion was found!

They are pathogens
Of earth's nastiest diseases.

However, they are everywhere,
And have two distinct states,
As no other proteins do,
One active and one inactive.

Without guidance from above,
They can switch states
And alter proteomic structure
Without changing DNA,

And then transmit their
New, infectious structure
To neighboring cells
With no transfer of genetic material.

Biology's sacred rules are violated!

In the brain, CREB proteins are
Sturdy enough to resist time,
They being virtually indestructible.

Yet, they have plasticity,
Being free of the genetic substrate,
To change their shapes,
Creating or erasing a memory.

When we think,
The neurotransmitters
Serotonin and dopamine
Are released by neurons,
Which switch the CPEB protein
Into its active state
By changing their very structure.

The activated CPEB marks
A specific dendritic branch
As a memory,
Recruiting the requisite mRNA
Needed to maintain
Long-term remembrance.

And, yet, prions have
An element of randomness
Built into their structure
Due to the inscrutable
Laws of protein folding
And stoichiometry,

Even becoming active
For no reason.

Ah, such contingency
Is just like Proust predicted:
The remembrance of things past
May not be the remembrance
Of things as they were.

Due to unpredictable and unstable prions,
We have some essential randomness,
For memory obeys nothing outside of itself.

GLOBAL WARMING ADAPTATION

When the environment changes,
Whether by man, the weather or climate,
Certain organisms who just happen to be resistant
Can then flourish, while the remainder perish,
Such as when man used pesticides to protect crops.

Whatever changes the planet
Alters the environment,
Be it man or Mother Nature;
Life has to go on from there.

Maybe if global warming came to pass
And it got really really hot,
Perhaps only certain organisms/humans
Who had some resistance would survive.

VIETNAM

Those were the days, of dissent;
The Army drafted; I was in,
Vietnam ongoing, objectors all about—
Jane Fonda, FTA rallies in Honolulu.

Took one look at the barracks,
Never registered therein;
Decided Waikiki was better,
A beach front by Diamond Head.

Saw a sign, "roommate wanted",
So I, the soldier lived and loved,
With she, the antiwar protester.

SENSE INTERPRETATIONS

It seems that our senses themselves
Are in direct contact
With what's really out there;
But we are not, for a face is
Painted upon it by the brain.

It is even perhaps an improved face,
Such as colors being distinct,
While the differences in wave frequencies
Might not be so or just some numbers
That would be hard to deal with directly.

We can tell, though,
From ourselves and/or from instruments
About an object's incidence angle of light,
Brightness, shape, intensity, amount, energy,
Texture and color (E/M wave frequencies),
Sound (air vibrations),
Odor (molecule shapes),
Taste(4-way vector of sweet,
Sour, bitter, salt), etc.;
So this is knowing something
Although not directly seeing it.

And, of course,
The building of devices that work
Shows that we know much
Of how what's out there
As reality operates.

A side thought is that
When a tree falls in a forest
With no one around there would
Be light waves but no light,
Air vibrations but no sound, etc.,
While in a night dream
There would be the opposite state—
Sound with no air vibrations,
Light with no waves, etc.;
So I suppose that
A dream is more direct
But not with so much use.

DARWIN'S THEORY

Darwin set the stage for understanding evolution
By noting on his 5-year *Beagle* boat trip
Such things as adaptive traits
In various finches on Easter Island
Or some place nearby.

Darwin and his pilot, a devout Christian,
Had many 'discussions'.

Mendel crossbred plants
To track the passing of traits
By dominant and recessive "genes"
That he didn't yet know everything about.

In the 1950's, genes were seen to operate
Via the 4 nucleotides on strands of DNA,
Further grouped into chromosomes (23).

Chimps have 24; however one of ours
Very much appears to be a fusion of 2.

Creationists stood firm, though,
Noting that the Bible said
That all the species were immutable,
Meaning that they were created as is,
And never changing.

While all of this science demonstrates
All the more our chemical nature,
Also declared impossible by some,
It is very much old hat,
And we see that the DNA chemical usage
Is a plus in that it is a digital code,
As in computers.

In computers/life,
An error can bring down
The whole system/organism,
And so there is error correction
Of bad transcriptions.

Computers have hardware error detectors
For bad parity and such,
And its software also does
Checking of its inputs.

The same for DNA, and so this error correction
Was discovered in genetics some time ago.

And, too, junk DNA was found not to be
So much pure junk but a combination
Of some things useful as well
As some things actually discarded—
Matching the fossil record.

We, too, find that genes may relate
That are not near each other in a linear way,
But near when a strand is folded in its usual way.

Also, that genes are not just for coding proteins
But for control of the overall system, too.

The evolution of millions of years, even billions,
Is recapitulated in the months of gestation,
From a few cells to a blastula to more,
All still taking place in water.

So, true science ever advances;
Religious dogma is stagnant;
It declares its knowledge base all at once
One of invisible guiding hands
And imaginary sources;
Quack science or
Pseudo science goes nowhere.

Creationists used to reject Evolution out of hand.

After having to accept it,
Some then claimed it as God's work,
But many others pointed to
Some gaps in the lines of species,
Ever pushing against,
Not having anything to push forward;

Then they tried irreducible complexity and failed;
The Pope excluded the organ of the mind from evolution,
As if that made evolution of the rest of the organ(s) OK;
Some even denied that evolution ever happened,
But then embraced it as Intelligent Design.

Some wished it to be banned from schools,
And failing that,
Pretended that Creation Science is a science,
Which it isn't,
Wishing for that to be taught
Alongside real science.

Some religious schools,
Being private, teach it in place of.

Science marches on.
GPS works because
Einstein's relativity works.
Newton's classical version
Still works at lower speeds.

Science embraces mystery
And the continued progress
To unravel it;
Dogma is a STOP SIGN,
In any field, saying "case closed".

So you agree that science shouldn't say
The case- against "God"- is closed?

Some do; some don't.
Science is open to proof, unlike dogma.
Dogma is a pronouncement without evidence,
Such as 'God did it'.

Science doesn't admit no-evidence entities
Into its body of knowledge.

Plus, the case can't even be opened
Since it is said that such evidence
Is not even conceivable;
Besides, only the natural has been found,
Nothing extra.

'God' is just a wish, albeit a beautiful one
In some of its rarer forms.

Evolution is fact, as seen in the DNA/fossil record;
God, though, is not fact, but only a belief.

DNA Information from God?

No, and yes, DNA is an information system,
As is also a being or a Being.

Positing a Being as the First is not useful.
All beings and Beings are systems of planning,
Learning, creativity, etc.,
And as such, are composite combinations;
So, they come later on, not first.

Look to the future for smarter and higher beings.
For something approaching the highest,

Look really far into the future.

Looking beneath and into the simpler past
Is the complete wrong direction
To find the ultimate complexity.

So, then, one must merely
Pronounce 'God', ungrounded.
And then what?
Stop? No longer wishing to explain it—
When previously there was
Such a need for explaining all?

There is still a great deal we have to learn!

Good, for that is a more open stance,
(Although we realize that there is still a hope
In the back of your mind that 'God did it'.)

MINDING THE BRAIN

Of course the brain is complex.
It doesn't just have
A million billion connections
Sitting around doing nothing.

The brain, mind you, and always the brain,
Even has a kind of really fast
Black & white type motion detector
Which allows you to duck from a swooping bird
Even before you've fully 'seen' the bird,
But, so what, for the brain does a lot of things,
Including monitoring
And attending to bodily functions.
As this would be but just a lot of noise;
It's part of the 95% of what the brain does
That does not bind into consciousness.

Now, what is the separated mind again
And what does it do that the brain doesn't do?
Also, what is the mind made of?
Is it just a higher part of the brain?
What's the definition?

WHAT IS CONSCIOUSNESS?

It is the brain experiencing representations
Of inner and outer states.

It is in consciousness
That the experiences of thought, actions,
And scenes that surface on the 'mind'
At any given moment from the brain's
Mostly underlying and subconscious doings
Of analysis are observed/witnesses/bound/presented.

This would suggest that the mechanics
Of the operation of consciousness
Are more or less the same for everyone.
Of course, the content of consciousness
Would differ greatly per one's differing experiences,
Emotions, learning and associations.

I am saying such that consciousness
Is the sea or stream (water)
In which the objects of experience
Appear and are 'seen'.

It is the sea in which we 'see'.
While consciousness is rightly referred to
As the subject 'I' in the English language
Since that's where things
Are felt, known, and told,
They are, as said,
Of the particular self
Of what one has become—
Again, of the brain.

It is akin to the visual system
Being more or less the same
In all of us that takes in the differing scenes.

That results of selves' doings
Appear 'last' in consciousness
Is a necessity since there
Must be time for the brain
To do its analysis, although this is often
Only a few hundred milliseconds at the least.

It is not that we are just tourists
Going along for the ride,
Though, although some of it is,
For the brain remembers the global experiences
Observed in consciousness for future usage.

There are some moo-goo people
Having some ungrounded notions
That consciousness comes first instead of later,
And this makes no sense at all,
For consciousness is a subject of experience,
And it is the experience that comes before,
Even before that it getting processed by senses,
Brain neurons, memories, associations, etc.,
This information working its way
(Mostly subconsciously) up to where
It can actually be witnessed and known—
In consciousness.

We don't notice any of the very short 'tape delay'.

If we are under anesthesia, for example,
The under workings may still occur,
But the results will not be known
Since consciousness has been shut off.

In casual conversation of the states of our being,
We distinguish between consciousness, mind,
Self, brain and awareness;
But, in the state beneath,
All is really occurring in the brain,
The smaller divisions of neuron assemblies
Competing for global attention
(Many little 'minds', perhaps),
Analyzing, forming words, thoughts, feelings,
Sense information or actions
That have achieved enough global resonance
To bind into what's presented to 'us'
In unity in consciousness,
Usually limited to just a very few experiences
At any one time.

This can be thought of as the 6th sense—
That of the brain's perception of itself.

Consciousness is not an object,
Except in the limited sense
That we can be aware of being aware,
Which puts a kind of sobering distance
Between the contents of the 'drama',
As if we are sitting way back in audience
Instead of the first row in front of the stage.

It is then that we should not
Fall hook, line and sinker
For all of our own thoughts

Merely because we thought of them,
But sometimes let the thought-parade
Just pass on by without necessarily latching on;
This is transcendence!

It can often be a subtle thing,
But let me first use an extreme example:
When I see a bad driver,
I don't get upset
But consider that they are too old,
Too young, too crazy, or just distracted;
However, some people might say
They'd like to kill the person.
Well, some actually might kill
But most seem not to,
Disregarding the thought.

Now, here's the tough case.
Suppose that the distinction
For disregarding any thought
Is not so obvious as
That one might go to jail if they don't.
Then, wrong thoughts can slip through,
Happen, and even roll on,
Some 'sins' and/or wrong ideas
Perhaps resulting, eventually.

If these scenarios were practiced
Over and over again
In a school course,
Then there could be more learning
Of the consequences of thoughts and actions.

KNOW TOE

Knowing a bit about the TOE
Lets us separate the wheat from the chaff
And not go off falling for numerous kinds
Of made-up and conflicting ideas,
All of which, paradoxically,
Claim to be true just by the saying so

A TOE has to explain how Everything works,
Both before and after the origin of the universe,
As that strengthens the whole scheme of the idea.

DRIVING LESSONS

We were vacationing
At a lake 500 miles away from Chicago
In upper Wisconsin in the North Woods.

I was 15 or so in the summer of '64
And learning to drive a '56 chevy
With a 3-speed stick shift.
This is the car whose taillight twisted
And lifted up to reveal the gas cap.

My father chose a cemetery for me to practice in.

I asked, "Why here?"

He said,
"Well, it's OK if you run over someone...
Because they're already dead."

He was a humorist, too, but that aside,
It was that my consciousness was very full
While learning to drive, my mind even having to
Have a special focus of awareness
On every little thing happening.

Eventually, driving became
Pretty much automatic,
And sometimes I have even wondered
How I got from here to there to here.

So, consciousness is necessary
For learning during the actual
And also for learning during
Imagining future events,
A time for internal doing
Without really doing,
Which extends even
To the nerve spindles
Throughout the body,
A way to actionize
Without movement.

We also have mirror neurons for learning
While watching others do things.

SEEING AND "SEEING"

I come upon an outdoor scene—
The Hanalei River Valley below and beyond.

Photons stream in;
The brain's visual systems I-IV analyze incidence angles,
Forms, brightness, intensity, texture and much more,
Combining and painting the face of them into unity.

The scene, actually seen only inside the head, as ever,
Is a projection based on the distances
Of the contents of the scene, etc.

Other brain systems 'know' the contents of the scene,
Such as trees, mountains and rainbows,
And their associations,
And this is added to the unity.

Memories may be associated,
And thoughts and feelings and qualia all arise
In the mind's eye,
Bound and observed in consciousness.

They appear in consciousness, not from it.

Consciousness could also be called
The mind's eye or perception.

Thoughts can be unbidden,
Coming out of the blue, so to speak,
Because what we call the 'will'
Acts at the subconscious level.

Only when thoughts surface on the mind
Do they become fully known,
For we are not always privy
To what analysis goes on beneath,
In the rest of the brain.

We can be surprised, in general,
By thoughts arriving,
And even more so
If they are forbidden thoughts,
Which are quickly banished
By some other part of the brain.

The trick is to apply some insight
Onto the less obvious cases
Of what may be spurious thoughts

It is often that simply because we thought of them
That we might give them absolute credibility
Without much pause.

UNSPECIAL

Just as there was no special time during eternity
(A universe could appear any time),
There is no privileged or special place
(Which must then be anywhere
And even everywhere).

Similarly, there is no specific amount of energy
(It cancels to zero on balance);
This all resolves any paradoxes.

As there can be nothing prior to the causeless
To define it, then either the specifics
Can't be any predetermined amount, place or time
Or they are random but potentially
Anywhere, any time.

The 'Who' is us, a complexity that came later.
The 'Why' is that there was not
A total lack of anything
Or this would still have been the case,
So, 'something' was the natural state.

The 'How' is that the balance
Couldn't stay balanced,
As the simple is always unstable;
It would have taken an outside force
To maintain the balance—and there wasn't any.

SOLIDITY?

The energy of a proton spinning
At the speed of light 10^{23} rpm,
Possibly having been compressed
From a larger volume into becoming a proton
May be the closest thing to what we think of
As solid on the micro level,
So, it could be that there is no real 'solidity'
But that the swirling energy makes it seem as so.

SOME MIND [BRAIN] DEFINITIONS:

'Mind' is not a separate entity from the brain,
But a shorthand notation
For the resources and processes of the brain.

'What's on your mind' is a shorthand
For what the brain has processed at the moment;
It is just a few things,
Not the brain's entire knowledge base.

'You're out of your mind'
Is about the doing of weird or crazy things.

Consciousness is a brain process
That culminates with the witnessing and knowing
About what's on the mind at the moment;
It doesn't float around by itself without a brain
To produce things on its own.
It can even be completely prevented
By some molecules of anesthesia to the brain.

Although the brain is fast,
Having 100 billion neurons with many connections,
The brain's analysis still takes some time,
Although not much; then the results arrive;
The brain has perceived
Its own momentary global state.

Not everything the brain does
Reaches consciousness;
Some that doesn't is for
The monitoring of bodily function
And some is of the unfinished analysis state,
Both of which would be a mumbo-jumbo
Of electric and chemical signals
If we could look in on it.

The self, too, is the same as the brain,
It's focus being all that we have become
And how we think and act.

Memory is in the brain, as well,
It finding symbols and
Their associations very quickly.

Senses are tied to nerves
That the brain monitors
And then post-processes,
To paint a more useful face upon.

Dreams, visions, and OBEs/NDEs
Also occur in the brain;
Full blown OBEs can be induced in the brain;
Poking the brain can cause memories
And visions to vividly appear.

Meditation is 'not what you think',
But causing a high degree of separation
Both from within and without.

Consciousness is not
Unconsciousness or subconsciousness.

A soul is some kind
Of invisible appendage supposed.

Consciousness would
Seem to be a process only,
A brain process in which
Some global kind of binding
Of a consensus of thought
(Or even a simpleton thought)
Becomes known/witnessed,
In that it surfaces into consciousness
As "what's on the mind" at the moment.

Before that the brain would
Have had to take some time
For analysis, a few hundred microseconds,
Speedily collapsing many scenarios of consequences
(Not necessarily meaning quantum-like).

One would not think that consciousness
Could just float around on its own;
Besides, it seems to be a subject, not an object,
Although we can be aware of being aware,
Which is not a bad method
For putting some sobering distance
Between the audience
And any trauma or drama onstage.

This is the normal case;
Some, people, though,
Who are perhaps
In need of antidepressants
Just seem to react emotionally
Without seeming to fully process the input
Or at least pause in the slightest
To allow for a more creative response.

Even though consciousness observes
The results of the subconscious 'will',
One would not just be a mere tourist
Along for the ride (maybe sometimes),
As surely the thought is remembered for future usage,
Which could even come in the next thought,
One that might even veto the first thought
("Let's not really kill that bad driver").

Note that self, mind, consciousness,
And all those kinds of words are still about the brain;
We just use them sometimes for a better focus.

THE UNIVERSAL ACID

As a boy in chemistry class, I wondered,
As did many,
About the following scenario often dreamt of:

*I mixed two compounds, which, unfortunately,
Produced the ultimate acid.
Nothing could contain it;
It quickly ate though the container,
The floor of the laboratory,
And then even all the way through the earth,
Eventually sloshing some poor sap in China.*

This, too is what happens to us,
Through education,
As our chemical-bio-electric nature
Is revealed to us,
Like some kind of giant shock,
After which we will never be the same again,
As perhaps some are now reeling from,
Well, maybe just a little bit.

The biochemical mush that is us,
When fully realized,
Leaves us stunned and astounded.
We grasp for what we once
Thought we were before,
But it eludes us
In the new light of learning.

*The universal acid of such knowledge
Eats through all superstitions,
Folk tales, and myths.*
Nothing can contain it.

BRAHMAN AGAIN

Mel, you never saw Brahman
And, so, obviously,
These tales were invented out of imagination;
The same for stating his nature and intentions
Or what he wants, etc.

Being a being or a Being takes a village,
That is, a composite system
Of mind for planning,
Knowing and creating, etc.,
So an Ultimate Complexity
Could not be first/fundamental, etc.,
Although it could be a smart alien.

Besides, beings happen not beneath and first
In the simple and tiny past,
But, beyond and later on as complexity.
You are looking in the complete wrong direction.
Someday, we may all join as one Being.

Nor was there any creation of the ground-state,
For nothing could have become of nothing,
And so the ground-state was natural and normal.

It was always here and still is.
As all of the conserved energies sum to zero,
The ground-state would be
The small residue of capability,
It balanced with pluses and minuses
Or able to differentiate itself into both,
Always emitting them in pairs.

So, energy doesn't come from nowhere
But is conserved.
(If there were a lack of anything
Then that would still have been
The case forever, but it wasn't.)

This ground-state was/is everything;
You can still see it or a semblance of it today,
For it never left,
As what produces the
Plus and minus virtual pairs
In the 'quantum foam'.
('Virtual' is a misnomer for real,
For they can stay real
If they don't go back in.)

NDES AND SUCH

If death is instant, more or less,
As some deaths may be,
Then one is really really dead and gone

Those reporting NDEs were not dead, obviously,
But only near death, since they are alive now.
It may sometimes take a while to fully die.

Near Death Experiences can be like OBEs
And/or with the extreme added effects
Of what dying contributes,
Such as the seeing of nonhuman beings
And/or tunnels
And bright lights of otherworldly scenes.

OBEs and NDEs are accepted,
Via many credible reports,
To be quite believable
As a realm of human experience;
However, they are just visions
Made by the brain.

NDE tunnels of light and such
Can be explained by neurology,
And OBE's by a condition
Called sleep paralysis;
They can also be induced,
Resulting in full blown episodes;
Neither, then, are proof of a beyond,
But of an altered brain state.

I found that I could awake
From dreams anytime
By clenching my whole body,
And so during an OBE
I luckily found myself
In a kind of halfway state
In which my dream-arms
Were seen to be fiddling
With the end-table stuff
While I could also see my real arms
Lying beside me unmoving.
Another time, I was able to
Keep some dream music
Playing after I awoke.

Sometimes a virtual dream reality
Cannot be told apart from the real;

I was so sure that I was out of my body,
But, one must also remember
That memory and imagination
Often picture scenes from above (try it).

It is also the case that people of different religions
See different religious figures during NDE's,
An indication that the phenomenon
Occurs within the mind, not without.

OBE's are easily induced by drugs.
The fact that there are receptor sites in the brain
For such artificially produced chemicals
Means that there are naturally produced chemicals
In the brain that, under certain circumstances
(The stress of an trauma or an accident, for example),
Can induce any or all of the experiences
Typically associated with an NDE or OBE.

NDEs are then nothing more than wild trips
Induced by the trauma of almost dying.
Lack of oxygen produces increased activity
Though disinhibition
Mental modes that
Give rise to consciousness.

What about the often reported experience
Of a tunnel in an NDE?
Well, the visual cortex is on the back of the brain
Where information from the retina is processed,
Lack of oxygen,
Plus the drugs generated at the time of dying,
Can interfere with the normal rate of firing
By nerve cells in this area.

When this occurs,
'Stripes' of neuronal activity
Move across the visual cortex,
Which is interpreted by the brain
As concentric rings or spirals.
These spirals may be "seen" as a tunnel.

Some saw scenes, yes,
But they were just made
By a mind in extremus;
They only seemed to go elsewhere.

NDE

THEISTIC EVOLUTION?

Actually, Poseidon causes plate tectonics,
And Ra initiates nuclear fusion in the sun.

Just joking, for yet no one today,
Except Pat Robertson types,
Appeals to these gods
To explain earthquakes or solar fusion;
Yet, it is proposed that a Higher Realm (called God)
Directs these things as well as all nature
And especially the mind of the brain (via a soul).

These thoughts are from the mythic ages,
And are somewhat still here today.

Some would go to the extremes
Of throwing all of science
Out the window as 'dogma';
Yet, their computers, devices,
And appliances run pretty well on this 'dogma'.

Let us then, ignore this stance as nonsense
And deal with the case
That evolution indeed happens
And that God directs it,
For that must become the fallback position.

For starters, evolution is not goal-oriented,
So we can discard the
(Biological teleological) argument
For the existence of God,
Which claims that postulating God is necessary
To account for purposiveness in nature.

Evolution is a *blind* watchmaker.

To review and elaborate more,
Theistic Evolution is the theological view
That God creates new species through evolution.

The advocates like to reserve
A special place for humans,
Separate from the animals.
But this is not
A scientifically justifiable stance
Given the many evolutionary
Predecessors of human beings.

Animals are 'brutalized' and humans humanized
To make the alleged gap as big as possible:
Humans are characterized
As the only creatures with reason,
Empathy, a (rich) emotional life, altruism,
Culture, identity, and language.
Yet all these characteristics
Have been observed
To a greater or lesser extent
In nonhuman animals,
Especially in other primates.

The history of the universe
Has been an unfolding
Of purely naturalistic processes.

The 'God hypothesis' provides
No additional explanatory value.
It is but a refuge of ignorance.

One who feels the need
To postulate a divine cause
Is left with the question
Of what caused God to exist.

Perhaps God does not need a cause;
But then why think
That the universe/stuff needs one?
So, it adds nothing.

Evolution is an immensely slow, wasteful,
Pitiless, and cruel process—
Hardly the most elegant process of creation
Open to a goal-oriented, omnipotent,
And benevolent God.

If humanity is the final goal of creation,
Whence the 3,500,000,000!
Years since the origin of life,
Or the 13.7 billion years since the Big Bang?
What is the point of this immense amount of time
If human beings and their world
Are the pinnacle of the Almighty's creation?

Does God cause mutations to direct evolution?
Well, they sure seem random,
Plus some are bad and many are neutral .

The vast majority of mutants
Are selectively neutral or negative
With regard to the evolution
And survival of *Homo sapiens*,
And, thus, their evolution is "wasteful"
If measured against the goal
Of producing human beings.

Such a wasteful process
Is hardly consonant
With a goal-oriented,
Omnipotent,
And omniscient God.

The case against theistic evolution continues...

The honorable Graybeard presiding,
Austin P. Torney continuing
As lawyer for the prosecution
(since his name contains the letters "attorney")...

"I call the recent family
Tree to the witness stand;
But, wait, oh my God,
There are some others, too,
Many of them extinct!"

The testimony:

There is no progressive trend in evolution
Toward the development of human beings;
Evolution can be seen as a huge tree
With many branching points,
Not a direct line to humans;
We are just a not-yet-extinct part of one
Of the very many branches
Of the enormous tree of life.

I now call upon the extinct.

Testimony:

What was the point
Of all these extinct animals,
If the goal of creation is man
And his surrounding nature?
To what purpose were the dinosaurs?
What was the point of the trilobites?
These groups of animals did not even contribute
To the origin of humans.

The development of life
Has been interrupted by
Innumerable extinctions,
Some with so many different plant
And animal species dying out
In the same time period
That they have been
Called mass extinctions.

Judge Graybeard,
Having worked for ten minutes straight now,
Calls a recess for a long lunch...

We see evolution differently then, Austin,
As the process of evolution seems most intent
On the continuation of 'life forms',
Though I will agree that it does not seem
To have a vested interest
In which forms are successful
In relation to which others.

True, no vested interest,
Those who were in a position
To 'adapt' and survive,
Were not necessarily 'nice'
(Could even be 'mean').

Still, I tend to wonder why all species
Of my experience offer a great struggle to survive,
When the alternative of doing nothing
Is by far the easier choice?

Survival is the brain's objective,
Sometimes even thinking
Its way into the afterlife.

Break time.

...

The Theistic Evolution trial resumes...

Judge Graybeard looks half-asleep,
But seems to be listening with one ear.

God is outside of time.

The prosecution answers and continues:

It has been suggested that God's mindset
Is very slow compared to the 'speedy' time

Of the operation of the universe,
And thus I submit
That God would not have been alert
Or responsive enough
To direct evolution through mutation.

The defense objected,
Stating that they would have to
Check with their Client on this,
The judge asking how long
Would this would take,
Noting that it took over 200 years
For a response to Haiti's pact with the Devil.

Oh, about a million years.

OK, we'll reconvene...
Wait! We can't wait that long...
So I'll allow the claim that God says
That He directs mutations.
Let the record show both opinions.

"Thank you, judge,
For our theory can adapt
To any and all turns of events.

The trial droned on...

Testimony:

Why would God create complete ecosystems
Only to have them virtually annihilated,
So that entirely different ecosystems
Would temporarily emerge in their place,
Only to meet the same fate, over and over again?

Had the asteroid which wiped out
The dinosaurs 65 million years ago
Missed the earth,
It's likely that our little branch
On the tree of life
Would never have developed,
Since the end of dinosaur dominance
Made it possible for our
Small mammal ancestors to flourish;
How are such chance contingencies
In the history of life compatible
With the alleged providence of a Creator?

Graybeard—the judge, was fully awake now
And was carving something out of a large block of wood,
Which was said not to be a boat or a cargo-cult thing;
But he still watched the proceedings with one eye.

Worse still, consider
The vast amount of suffering
Needed to secure our existence
Through natural selection;
The environment "selects"
Those organisms best adapted to it
Not the most even-tempered ones.

Consequently, numerous predatory creatures
Have evolved which regularly inflict suffering
On prey and host animals.

The screw-worm fly (*Cochliomyia hominivorax*),
For instance, lays its eggs in the wounds or eyes
Of mammals (including humans),
Causing any wounds to wide
When the eggs hatch
And the larva eat
The surrounding tissue.
This attracts more congeners,
Further widening the wounds.

Untreated, such parasitism
Often leads to a gruesome death.
Or consider the human immunodeficiency virus (HIV)
Which causes autoimmune deficiency syndrome (AIDS);
Is a great evolutionary success one which creates
Immense suffering among human beings?

It was now getting near 3 PM,
The judge announcing,
"That's it for today;
Let's meet again sometime after my vacation.
...

"The mind is like a man in a rowboat."

A few days ago there were 50-foot waves
On the North Shore of Oahu.
Some ants looking like people surfed on them
Or at least the wave remnants,
Enjoying the ecstasy,
And then were ground up
Into the sand—the agony;
A guy in a rowboat fished them out.

I can see a tree with my physical eye
Because that tree is embodied in matter,
But, to conceptualize that tree
My mind goes into universal concepts of meaning
Which have never taken embodiment in matter,
For they do not require physical presence before my eye.

So, you see, concepts are not acts of a bodily organ
Such as the brain or they would exist in matter;
Conceptual thought is an immaterial power
Which we use to form concepts of meaning.
Utilizing that power does not require
Any physical sense or organs.

Judge: Immaterial,
Since the brain is an organ.

...

The Theistic Evolution case resumes:

Attorney: I'd like to address the tree.

Judge: Proceed.

Attorney: Hello tree.

Judge: Ha-ha.

Attorney: The tree, as out there,
Is a bunch of waves,
The photons carrying the 'visuals',
The air-vibrations passing on the 'sounds',
The molecules transmitting 'odour'
By their shapes, etc.

I use quotes to show that these transformations
Are fully made later on by the brain;
If there is no brain around,
Then there are just the waves emanating.

While our senses are absolutely
In direct contact with the waves/particles
That are out there,
We don't have awareness at this level,
Plus, the direct jumble of waves
All interfering with each other
Might not reveal anything much right off the bat.

So, the brain proceeds to process the information
With its many modules and subsystems
Through higher and higher levels,
Finding edges, intensity,
Color, and distance for vision,
Detecting molecules shapes for smell,
Interpreting air waves as sound, etc.,
As often much more detailed elsewhere,
Until the tree is seen, smelled, heard, and so forth
As the final perception of the tree
With its qualities within the head.

There is no dividing line
Where the brain says
"I can do no more"
And hands it off to
Some nonphysical realm,
For it has already done it all.

This includes the brain's memory coming along
And knowing what a tree is as a whole and its parts,
Associations arising, such as the old tree house
Or that leaves have chlorophyll
And fall in the autumn, etc.,
And then more associations
Upon those associations.

The brain is the lifeboat navigating
And re-cognizing the waves of reality,
Painting a useful face upon the waters.

All is ever in the brain as a representation,
The tree never being directly known as matter,
Not even in the first place.

Judge: OK, back to evolution testimony:

Immense suffering,
Like wasteful "trial and error,"
Is not incidental,
But is inherent to the process of evolution.
And it does not sit well with the notion
That evolution has been set up
Or directed by a loving God.

The theistic retort that
"God moves in mysterious ways"
Goes well beyond the evidence
From evolutionary biology,

Not to mention that it is a kind of excuse
For very poor, sometimes seemingly near insane,
Ways of accomplishing things.

There is a far simpler
And elegant explanation
For that evidence:
There is no divine will
To grope at in the dark,
Just the indifferent, pitiless,
And naturalistic forces of evolution.

Since evolution is a slow, wasteful,
And brutal process,
Prima facie it is not the way in which
A goal-oriented, omnipotent, omniscient,
And loving God would choose to create the world.
Thus a naturalistic explanation
For the origin of all species,
Including *Homo sapiens,*
Is more plausible than a theistic one.

Judge: Nap time.

A-VOID

A void is not there,
And so it must be avoided
So as not to step into nowhere.

The void that is not there
Is emptiness, nothingness,
Nullity, blankness, and a vacuity.

So, it is, compared to what is there,
A gap, cavity, chasm, abyss,
Gulf, pit, and a hole.

One must invalidate, annul, nullify,
Negate, quash, cancel, countermand, repeal,
Revoke, rescind, retract, withdraw,
Reverse, undo, and abolish the void,
Abrogating its existence, for it is not to be.

One cannot really make
So much ado about nothing.

THE THEORY OF EVERYTHING:
ALL THE WAY UP

STABILITY MARCHES UPWARDS

Everything
Becomes less complicated and smaller
As we delve downward
From the complex toward the simpler
As it must;

And so then does the simplicity of
The Theory of Everything,
For complexity lies
At the other end of the spectrum,
Where we are.

THE SEARCH

I'll follow every single avenue,
Whether it's brightly lit or a dark alley,
Exploring one-ways, no-ways, and dead-ends
Until cornered where the truth is hiding.

The TOE must not only encompass
The unification of the forces,
Which is the GUT
(The Grand Unification theory),
But it must also demonstrate
Why anything at all exists and how it did so,
And, furthermore, how that ties in
To what we humans have become up to now,
How we operate, and so forth.

"Why does anything exist at all?"
Is much akin to the 'great' philosophical question
Of *"Why is there something instead of nothing?"*

Both questions are stated backwards,
As if 'something' had to be made,
Presumably from an absolute Nothing;
This is not the case,
For causes cannot forever precede causes,
Nor can a total Nothing ever produce anything,
For 'it' has no existence,
Meaning that it is not even 'there'.

So, the eternal causeless 'something' must then
Be the normal and natural state of affairs, for,

1.
If there were a total lack of 'something'
Then this would still be the case,
As there would continue to be no existence.

2.
A chain of ever caused 'somethings'
Would be a never-ending infinite regress.

3.
Nothing, obviously, couldn't exist or persist.
See it? No; it's not there.

So, what is this 'something'?

1.
It is simple 'thing',
For the complex ever
Always has simpler parts
From which it is constituted
As we look downward.

2.
What is as simple as it gets?
Nothing, but that is not there.

3.
Then what is the next simplest 'thing'?
Well, it must be a near 'nothing'.

4.
Was 'it' designed or reasoned forth?
No, as it was causeless—
Being the 'first' and always.

5.
So, we don't expect a blueprint of it
Since it's an undefined chaos of random disorder.

Well, (5) is, of course, hard to swallow,
As we always wish for definition;
However, the unordered must be the answer,
For there was nothing prior to it
To order it or to reason/design it forth.

*So, then, 'something' exists
Because there was no alternative?*

Yes, that is the 'Why' of existence,
For 'something' is the natural and normal state.

As usual, like all simple things,
The near 'nothing' is unstable,
For simple things ever go through phase change
And many even combine into the more complex.
This might even apply to a total Nothing
If it could even 'try' to be.

How do we know that
The 'something' is a near 'nothing'?

1.
The negative energy of gravity
Cancels out the positive energy of matter,
Leaving only the residue of
The near 'nothing'/capability.

2.
Charge and angular momentum
Also add to near zero.

3.
Thus, again, the residue must be tiny,
And, of course, this makes sense,
For it is seen that the universe
Is even still expanding.

What is the near 'nothing' exactly?

1.
There is no 'exactly', due to chaos,
Perhaps wavering between existence and non;
It preceded real form, space,
The quantum (maybe, or it is it), and laws;
However, it could be considered
As the quantum realm
Or its semblance,
In which the near 'nothing'
Is the quantum fluctuation,
Also called quantum tunneling
Or quantum uncertainty.

2.
It creates real particles,
Some of which are stable and enduring
Or at least capable of combining
Into the larger real
If they don't fall back in.

Ok, but let's back up.
This 'something' was always there

And was never created?
True; it had to be.
Not a thing can become of nothing.

Yes, but no creation means no Creator.

True. We are free to be,
And also free from that superstition.

And the 'How' was that anything could become
Of this 'chaos' of possibility,
Anywhere, any time, any size,
Such as a universe?

Yes, for we even see it
In the disordered quantum realm.
We measure atoms,
Each time getting a different answer.
Radioactive decay just happens
Whenever, unpredictably;
We even now see the superpositioning of electrons
In green sulfur bacteria, via fermo-lasers,
As these electrons in superposition
Locate the most efficient path for photosynthesis.

OK, then what is the 'Where'
That we human mammals utilize?

Space is the 'Where'.
It is a place for the 'What', which is matter.

Then whence derives our 'Then' and 'When',
The past and the future, with the 'Now' in-between?

The movement of appearances,
That being of the 'What' passing through the 'Where',
Gives rise to the notion of the past, present, and future,
A correspondence that is retained in memory
And in imagination's projective powers.

Huh?

Remembrance, or memory,
Is the past space of the 'Then' and the 'Where',
Whereas, history is the past matter
Of the 'Then' and the 'What',
Remembrance and history
Ever combining into learning.

The future space is of
The 'Where' and the 'When,
This being your hopes and wishes;
Whereas, the future matter
Is the 'What' and the 'When'
That makes the actual progression to progress;
Wishes and progression combine into vision.

Yes, vision, but what do
History and progression lead to?

A change-in-structure.

How about the result
Of remembrance and wishes?

They lead to a change-in-outlook.

OK, we're building up
Some more complex shells here,
But all stemming from the
Simpler movement of what appears;
So, what becomes of learning
And a change-of-outlook combined?

Your [new] direction in life.

And of learning and a change-in-structure uniting?

Creating.

OK, and of a change-of-outlook and a vision?

Growth.

And of vision and a change-of-structure?

Planning.

And, finally, what of direction,
Growing, planning and creating?

That's your being—the 'Who'.

Alright, we've seen the beginnings
Reflected in our being,
But what led from quarks
Or whatever to our complexity,
Taking 13 billion years?

Quarks formed into photons and neutrons
Which then collected into stars via gravity
Which then generated the lower elements.

Some stars exploded as supernovae
Which spewed more of the simple,
And even the higher elements into space,
Which then formed molecules that led to cells
That eventually combined into life,
Via the dispersion of energy,
That brought forth
The brain and consciousness.

That's the easy part of the TOE,
For all that's well known;
Of course, planets formed
And bacteria exuded oxygen
Into our atmosphere.

Another few billion years and we came along,
Through evolution, thanks in part to the dinosaurs
And 90% of all the species dying out,
Not to mention that two chromosomes fused,
Having us branch off from the proto-chimps.

Why did this all take so long?

Death was the only chooser,
Plus complexity just takes time.

Well, lucky us.
So what the heck should we do,
Thrust into this life as we are, without asking?

You are truly free to make your own meaning
Out of life's happenstance.

I feel liberated.

THE FACT OF EVOLUTION

Evolution simply means
That a species undergoes
Genetic change over time,
The differences based
On changes to DNA.

All was mutable.
The record is in the fossils and the DNA.

WANTING TO BE MORE THAN IT WAS

Perhaps the fact that death exists
Equates to "choosing" to be more
Than it was "before"—
Since it is only death
That sifts the best from the rest,
Via natural selection.

Some mutations are bad
Enough to cause death,
And many that don't,
Don't much matter right away
And can eventually selected in or out
By changes in the environment.

When the industrial revolution
Covered plants and trees
In London with black soot,
The white moths were
More often seen and eaten
Than the darker colored moths—
And so they pretty much went away.

Thus, death does away with those
Not "choosing to be more than they were before".

Death separates [chooses]
The wise from the silly,
The pointed from the pointless.

At the molecular level, too,
Some molecules arrange into assemblies
More composite and complex, and some don't.

REAL = ENDURING

Real particles are excitations
In quantum fields
That have a very useful degree
Of permanence
And can be observed.

As for virtual particles,
They can be turned into real ones
By supplying enough energy.

I AM A DREAM CHARACTER

So therefore, I am a 'thought' .....
So what is it that knows this 'thought'?
It is my conscious awareness that knows it. — Mel

That would really be Brahman's thoughts,
As dream images have
No thoughts unto themselves,
Or even the 'I' of a self,
But perhaps Brahman somehow
Localized a self into an actual you.

The other thing about consciousness
Injecting pictures and videos of pseudo-life
To... somewhere?...
Would be that our retinas
And all the visual processing behind it
In the brain would not really be doing anything
And so, then, we'd have to
Wonder why they evolved,
Along with the same for
All the other senses, memory, etc.,
And all the other bodily systems,
Not to mention the entire earth
And the universe being here.

THE LOST SCHOOL

I looked for the ancient mystery school;
However, its whereabouts remain a great mystery!

I looked deep into the unified field
For the hidden knowledge
Of our genetic encoding of 7-fold consciousness,
But, it was really too well hidden!

Then I looked for my tennis shoes
And there they were, right under the bed!

— THE END —

www.ingramcontent.com/pod-product-compliance
Lightning Source LLC
Chambersburg PA
CBHW071405170526
45165CB00001B/185